Pagans In My Blood

Pagans In My Blood

John Magor

ISBN 0-88839-291-5
Copyright © 1992 John Magor

Cataloging in Publication Data
Magor, John, 1915-
   Pagans in my blood

   ISBN 0-88839-291-5

   I. Title.
PS8576.A36P3 1992   C813'.54   C91-091780-9
PR9199.3.M346P3 1992

All rights reserved. No part of this publication may be reproduced, stored in a retrieval system or transmitted, in any form or by any means, electronic, mechanical, photocopying, recording or otherwise, without the prior written permission of Hancock House Publishers.
Printed in Hong Kong

Published simultaneously in Canada and the United States by

HANCOCK HOUSE PUBLISHERS LTD.
19313 Zero Avenue, Surrey, B.C. V3S 5J9

HANCOCK HOUSE PUBLISHERS
1431 Harrison Avenue, Blaine, WA 98230

Contents

	Dedication 6
	Acknowledgments 7
1	Wanted: A Home 9
2	Like Mother, Like Son 21
3	Enki and RJ 30
	Lincoln Magor 41
	Joanne Hammond 43
4	A Journey into Myth and Nationhood 48
5	Bell Jars 63
6	One "Enchanted?" Evening 78
7	Polish Mysteries 87
8	Wolski Meets Aliens 92
	Full Particulars of the Incident 94
9	Men or "Strangebeasts?" 99
	Look and Behavior of the Creatures 99
	A Friend's View of Wolski 102
10	Hell is for Real! 104
11	A Pair of Queens 110
	Index 114

Dedication

This book is written in memory of Kate Seredy whose enchanting tales of *The White Stag, The Good Master,* and others awakened thoughts of our beautiful ancestral homeland, Hungary, which my sister Felicia, and I finally came to know.

—John Magor

Acknowledgments

For helping to develop Pagans into a book, thanks go to:

David Hancock, Publisher, a quiet professional.

Barbara Buske, Office Manager, she was indispensable and imparted confidence to others.

Barbara and Wayne Buske, for generous use of their picture maker and other help.

Group in Poland for meeting and other help.

Zecharia Sitchin for excerpts from *The Earth Chronicles*.

Emil Lengyl for excerpts from *1,000 Years of Hungary*.

Kate Seredy for excerpts and illustrations from *The White Stag*.

Alexander Hislop for excerpts and illustration from *The Two Babylons*.

Gerhard Helm for excerpts from *The Celts*.

Andy Roberts for excerpts from the *MUFON Journal*.

John Robert Colombo for support, advice, and excerpts from *Mysterious Encounters*.

Penguin Classics for excerpts from *The Discovery and Conquest of New Spain* by Bernal Diaz del Castillo.

Kazmer Ujvarosy for passages from religious and Magyar literature.

Graham Conway for proofreading, UFO expertise, and being a good "old friend."

Mike Strainic, whose encouragement in UFO matters was by itself a welcome contribution.

Felicia Magor for illustrations, commentary, and company in travel.

Lorna Lake for help in production and her book expertise.

June Villalon for being ever my helpful sister.

Lincoln Magor for encyclopedic and other help.

Lillian Magor for patience, advice, and her care for a "writing" husband.

All Magors who shared with me family eye-openers we never knew about.

"Those who want to hear the voice of pagan gods . . . can follow the thread on the pages of this book. It is a fragile thread; it cannot bear the weight of facts and dates."
—Excerpt and Illustration from *The White Stag* by Kate Seredy.

1

Wanted: A Home

As legend partly tells the story, about four thousand years ago in Asia, two young men and a few "moon maidens" embarked on an extraordinary venture. It was never surpassed for its singleminded dedication and its significance in the small world of its time. Also it survives to this day on a heroic level of its own. For what this assorted party was daring to accomplish, and did so generations later, was to find a home for the new nation of Hungary.

Legend has it that at the start a White Stag appeared from the forest and led the way, but the rest was hard history. Before the little band reached the shores of the River Danube where their journey ended, they had crossed two towering mountain ranges and most of a desolate continent.

My reason for wishing to write about this epic venture, while faltering an the thought of it, is simply that one of the men was named Magor. With his brother Hunor, he was leader of the group, and his name has been perpetuated ever since in its evolved tribal form of Magyar.[1]

Despite the name relationship, however, I might have shirked the task of writing the story had I not discovered a small book published in 1916 and written by Rev. Alexander Hislop[2] whose religious animosities almost destroy the value of his work. To me, however, the book is a rare find because it contains the only account I have ever read of the exceptional woman concerned who was so wonderful to behold that to writers and painters she became known as the Madonna—and it was she who may have been responsible for reintroducing the name of Magor into our family, adopting it from the earlier ancestor. This could account for my sense of special relationship with her and the strength of my instinctive belief that I know her as a living person who was "born again."

Although obviously she entered the scene long after Magor and Hunor and, with all the artfulness she showed, was quite unlike the original "moon maidens,[3]" it was this Madonna who was clearly responsible for assembling the reins of history. Events then took a form which made possible the founding of a nation. Why her name, Semiramis, remains so obscure in that history baffles me, and I can only suppose it was because she was overshadowed at first by her husband who, though enormously well known in his own right, was a black man from the African Kingdom of Kush. Moreover his name became attached with sinister implications that made it

---

1. Serendy, Kate. *The White Stag.* Lengyl, Emil. *1,000 Years of Hungary.*
2. *The Two Babylons.*
3. In the dawn of Christianity moon maidens commanded deep respect as children of their expected God.

even more difficult for writers to adorn him with bland heroics of the textbook kind.

He was Nimrod, and it is only because of the little book I found that I can offer a peephole past the infamies that have muddied his name. So I have started my story with events developing past Nimrod's time of renown and progressing to the schemes of Semiramis for whom I have an affection, despite what I know would be the disapproval of The Rev. Alexander Hislop.

In doing his best to give a date for Semiramis, the sometimes vague author wrote (in a footnote!): "The age of Ninus (Nimrod) synchronized with that of Abraham, who was born B.C. 1996. . . . Semiramis is said to have survived her husband forty-two years."

Separately he added: "Now the period when Semiramis lived—a period when the patriarchal faith was still fresh in the minds of men, when Shem was still alive—made it hazardous all at once and publicly to set up such a system as was inaugurated by the Babylonian queen."

Although Hislop disliked everything that the heathens Nimrod and Semiramis stood for, he was too nimble a scholar to miss the chance of commenting historically on Nimrod's color. Thus he found a parallel for Nimrod's case even though it meant dipping into mythology and he wrote:

> . . . it must be evident that the figures in the upper compartment (of an ancient woodcut in the book) represent the fire-god, or Sun divinity: and what is worthy of special note is that these fire-gods are *black* [author's italics] the color thereby identifying them with the Aethiopian or *black* Phaethon. . . .

'All the faces in his [the artist's] engraving are quite black,' he adds in a footnote. 'In India the infant Crishna, in the arms of the goddess Dovaki, is represented with the woolly hair and marked features of the negro or African race.'

However, in the middle of history herself, Semiramis was not looking for scholarly footnotes. What she wanted with vision far ahead of her time was a laundered history, and the solution she found certainly removed any dark spots. Presumably only she was any the wiser about how it was done. Religious fantasy took care of the rest.

As Hislop explained, the very practices followed in Babylon at the time of Nimrod's death played squarely into her hands. These practices were known as the "Mysteries" and usually involved a priest speaking behind curtains in the name of the newly departed. "What could be too wonderful or incredible to be believed?" Hislop asked. " . . . the whole system of the secret Mysteries of Babylon was intended to glorify a dead man; and when once the worship of one dead was established, the worship of others was sure to follow."

Here the author presumably meant others who were still alive for it was into this phase of the Mysteries that Semiramis deftly slipped herself and a mysteriously acquired infant of white complexion, who would henceforth be known as The Son. So that suddenly her followers had a Holy Trinity to worship.

With obvious relish Hislop enlarged on his description of human failings which made this enormous hoax possible.

> The scheme, skillfully formed, took effect. Semiramis gained glory from her dead and deified husband; and in the course of time both of them under the names of Rhea and Nin, or "Goddess Mother and Son," were worshipped with an enthusiasm that was incredible, and their images were everywhere set up and adored. Wherever the negro aspect of Nimrod was found an obstacle to his

worship, this was very easily obviated. According to the Chaldean doctrine of the transmigration of souls, all that was needed was just to teach that Ninus (Nimrod) had reappeared in the person of a posthumous son, of a fair complexion, supernaturally borne by his widowed wife after the father had gone to glory. As the licentious and dissolute life of Semiramis gave her many children, for whom no ostensible father on earth would be alleged, a plea like this would at once sanctify sin, and enable her to meet the feelings of those who might have no fancy to bow down before a negro divinity. . . .

In the single statement that Semiramis had "many children" a new field of questions is opened up in tracing our ancestral line. For immediately we have the elusive mystery of why all the Magors from that period were of fair complexion despite the bold and ubiquitous presence of the man from Kush. We grant the bible fails to list any children at all for Nimrod and assume he may have been sterile. But the assumption falters when we consider the family tree continued to flourish before and after his death, and a large family more than filled the emptiness that an ancestor without children would have left.

If Nimrod with all his masterful qualities did not do his part in replenishing the family ranks, who did? Obviously for the answer we must now look directly at the attractive, artful, and above all fertile Semiramis, and we suppose she had the help of a coterie of consorts who would then have infiltrated the Magyar ranks as candidate fathers.

Still, the beauty of Semiramis was such that she never seemed in any danger of having to take second place in the plot she had helped arrange. On the contrary, she rose so easily above any such difficulties that to her wor-

Also known as Semiramis, was the Goddess Diana of Ephesus shown in woodcut with her pagan trappings. The tower atop her head was a symbol of Babylon founded by her and Nimrod.
—Woodcut from *The Two Babylons* by Rev. Alexander Hislop.

shipful followers she became virtually a saint. Of her ascent to that lofty realm, Hislop wrote:

> There was an express promise that necessarily led mankind to expect that, at some time or other, the Son of God, in amazing condescension, should appear in this world as the Son of man. But there was no promise whatever, or the least shadow of a promise ever to lead anyone to anticipate that a *woman* should ever be invested with attributes that should *raise* her to a level with Divinity. It is in the last degree improbable, therefore, that when Semiramis, as the mother, was first exhibited with the child in her arms, it should be intended to give divine honors to her. She was doubtless used chiefly as a pedestal for the upholding of the divine Son, and holding him forth to the adoration of mankind; and glory enough would be counted for her, alone of all the daughters of Eve, to have given birth to the promised seed, the world's only hope. But while this, no doubt, was the design, it is a plain principle in all idolatries that that which most appeals to the senses must make the most powerful impression.
>
> Now the Son, even in his new incarnation, when Nimrod was believed to have reappeared in a fairer form, was exhibited merely as a child, without any particular attraction; while the mother in whose arms he was, was set off with all the art of painting and sculpture, as invested with much of that extraordinary beauty which in reality belonged to her.
>
> The beauty of Semiramis is said on one occasion to have quelled a rising rebellion among her subjects on her sudden appearance among them; and it is recorded that the memory of the admiration excited in them by her appearance on that occasion was perpetuated by a statue erected in Babylon, representing her in the guise in which

she had fascinated them so much. This Babylonian queen was not merely in *character* coincident with the Aphrodite of Greece and the Venus of Rome, but was in point of fact, the historical original of that goddess that by the ancient world was regarded as the very embodiment of everything attractive in female form, and the perfection of female beauty. . . .

However, being so preoccupied with his fetish about idolatries, Hislop missed a vital point which, if observed, might have given him a more significant status as a historian and writer of his time. He failed to note that the infant in the arms of this beautiful woman might have been truly hers. To him it was all part of a calculated pretence for the greater glory of this already famous "Madonna."

Perhaps his oversight was forgivable, as to this day, no one else of record has made that observation either. But surely it was strange to others that the Son seemed to be playing his part as successfully as the Mother. At any rate, nothing in Hislop's somewhat skeptical account suggests anything was amiss in the infant's behavior during long hours of posing and public inspection. Instead, to judge from his smiling pictures, the child behaved quite happily as if in the arms of its real mother. The artists, in turn, apparently had no trouble in depicting the mother with an adoring look. So to me anyway it appears likely she was handling, and perhaps nursing, a baby of her own.

While appreciating what this exceptional woman did in saving us Magyars from what could literally have been a trip into darkness if Nimrod was not sterile as we suppose, we are still left with the formidable problem of deciding who the father was. Was it some charming but nameless courtier? Was it a chance acquaintance, or was it an ambitious accomplice who thought in this way he could win a place in the royal circle?

Any of those answers might suffice but the finger of guilt . . . er . . . approval must surely point first of all to a man well known to Semiramis. As we have seen, she planned her moves carefully. Having progressed this far, she would not throw it all away on a comparative stranger, no matter how charming. Her whole future in this duplicity would depend on a man she knew intimately who could be counted on to help her in foreseeable ways. Also he would have to be a man of leadership quality. One day he might be at her side in deciding the future of a nation still unnamed or even unformed.

Such a man might have been Bendeguz, father of Attila, leader of the tribes now making an exodus from Asia and one who would predictably be patriarch of the nation whose founding and place of settlement they sought.

One night while writing this I had a dream about the people in these pages. It was so vivid and lingering I thought afterward something more was involved than the effect of working on this story. Maybe by some psychic occurrence I was seeing those events through the ancient eyes of my ancestor Magor, who may have been there. In any case what I saw in the astonishing ceremonial pageantry before me—and somehow I knew instantly what was happening—was the marriage of Bendeguz, who had a curiously sedate even statesmanlike appearance, to Semiramis. She was dressed in flowing white like a bride of today and was surely the most beautiful woman ever seen by a man in or out of his dreams.

The ruthless young queen reproves an admirer.
—Felicia Magor

If a child was there, I did not see him but in my mind I think I have completed the picture. Since history records that Bendeguz fathered Attila,[4] what I saw may

---

4. Kate Seredy. *The White Stag.*

have been a glimpse of events soon before the boy's birth. If so I may have seen not only a beautiful historic moment but one of incredible irony because, as we know, history records that Attila became the terror of his enemies so that his name still is one of the most fearsome ever heard.

The irony is that with the smallest of imaginary touches, this ferocious warrior may have been the one pictured in his infancy as our worshipped little Saviour.

The ghosts of the queen and her subjects haunt "the watershed."

—Felicia Magor

2

Like Mother, Like Son

One of the most spectacular natural features of North America is a gigantic gash, 50 miles wide in places, running from the northern wilds of British Columbia down into Montana. It is the Rocky Mountain Trench, source of the great Columbia River that escapes northward from the valley before doing a U-turn to find its Pacific outlet far to the south in Oregon.

Fed from each side by watershed streams, the Trench was the basin of an inland sea half a billion years ago and today its rock walls are encrusted with relics of marine life strikingly out of context with this towering countryside so distant from the sea.

A landscape of magnificent contrast from its fertile bed to its snow-brushed peaks, the Trench looks like a playground of gods. And perhaps it truly is, for here there is a seemingly endless record of strange visitors from the sky.

—*Canadian UFO Report,* Vol. 2, No. 3.

Maybe the visitors come for the concealment the Trench offers, but for these will-of-the-wisp figures the reason is likely to be more subtle. It may have something to do with the mere appearance of this place. For here there is a constant play of water which, coupled with other changing effects and a sense of secrecy lurking there, creates an atmosphere which the gods may find to their liking. While Semiramis was not a deity, she was in the minds of her followers close to being one, and it is not difficult to imagine her spirit having an affinity with a god-like place such as this where ancestral souls might freely assemble.

The thought becomes all the more pertinent if we see Semiramis herself as the personification of a watershed, or an elevation around which other currents flow, for she suited that imagined part to a remarkable degree. Let's look at it.

First there was the phenomenon of Attila, her son as we believe he was. With parents like Bendeguz and Semiramis, Attila had promise of becoming an exceptional figure in history. Not only did he keep that promise but, to an uncanny degree, gave us an even more striking example than his mother of our watershed metaphor. In other words, he was at the center of even more forces than she, and in his case they were all terrifying.

Again we turn to Emil Lengyl. In *1,000 Years of Hungary* he writes of the mobility of the Magyar warriors and the terror they spread as they attacked through the Vienna Gap where the Danube channeled through a fringe of the Carpathian Mountains.

> The Occident has not yet forgotten the Huns, inhuman monsters in human form, under the leadership of Attila, whom the world was to know as the Scourge of God.... They were a people on horseback, moving under

the irresistible momentum of their march across the plains. Leaving behind holding forces, the Hungarian forces erupted from the Vienna Gap and turned their faces west. Beyond the Gap they branched off in different directions.... In one year the Magyar marauders were in what is northern Italy today and the next year they were in Burgundy, much farther west. In other years they followed the Danube and the Rhine, reaching rich

With Percell mountains at left and Rockies at right, the Rocky Mountain Trench stretches north through British Columbia like a ditch dug by giants, offering what seems to be a favored area for UFOs.

—This is a B.C. Government Air Photograph courtesy of the Department of Lands, Forests & Water

lands which they traversed and laid waste. In one of these years they were making a race to Rome, while in another campaign they swept along the Adriatic coast of the Apennine Peninsula, heading toward the distant south. They scored victories and established tributary regimes.

Like water thundering down from another level and finding its way through new channels, it was orderly confusion. And in that state of contradictions again we have an invitation to extend our metaphor, and again it is a case of like mother, like son. For with all their extreme ways, both had a sense of exactness.

With Attila this was particularly evident in his brilliance for precise timing. He gave an example in making his thrust through the Vienna Gap, a feat that finally led to the Magyars' almost total victory in central Europe. Lengyl writes in *1,000 Years of Hungary:*

> The Magyars appeared in Mid-Danubia at just the right time, when it was not under a strong central rule and marginal regions of transmountain realms were engaged in suicidal jurisdictional disputes. The Mid-Danubia was thus a typical power vacuum. The Magyars destroyed Great Moravia in a series of wars, routed the Khazars and Vlachs, chased the Bulgarians further south . . . the Danube valley was a marginal land and the Magyars moved into it with alacrity.

In an entirely different way but with the same penchant for neatness and leaving her mark like the strong stream of influence she was, Semiramis continued her religious practices at which she had already proved so adept. In doing so, she left for posterity a mark of the pagan traditions that governed her life. She did this while bearing her alternate name of Cybele, and Hislop records the period:

> . . . every scholar knows that when the worship of

... every scholar knows that when the worship of Cybele, the Babylonian goddess, was introduced into Pagan Rome, it was introduced in its primitive form with its celibate clergy. When the Pope appropriated so much that was peculiar to the worship of that goddess, he introduced into the priesthood the binding obligation of celibacy. ...

These celibate priests have all a certain mark set upon them at their ordination; and that is the clerical tonsure. The tonsure is the first part of the ceremony of ordination; and it is held to be the most important element in connection with the orders of the Romish clergy. When, after long contendings, the Picts were at last brought to submit to the Bishop of Rome, the acceptance of this tonsure as the tonsure of St. Peter on the part of the clergy was the visible symbol of that submission. ...

Now, as Rome set so much importance on this tonsure, let it be asked what was the meaning of it? It was the visible inauguration of those who submitted to it as the priests of Bacchus. This tonsure cannot have the slightest pretence to Christian authority. It was indeed the 'tonsure of Peter,' but not the Peter of Galilee, but the Chaldean 'Peter' of the Mysteries. ...

The high antiquity of this tonsure may be seen from the enactment of the Mosaic law against it. The Jewish priests were expressly forbidden to make any baldness upon their heads, which sufficiently shows that, even as early as the time of Moses, the 'shaved head' had been already introduced.

Although it took a giant leap from the Columbia River to the Danube to draw both Semiramis and Attila into our watershed we find that curiously the metaphor we have used lands us in a place still closer to its original definition. For there are few places in the world closer than the Vienna Gap to the physical requirements of a watershed, as here we have country serving as a drain-off for the Danube after its waters course down from Alpine heights. Also it has picked up mysteries from the dark realm of Transylvania past which it flows just as the Columbia picks up phantoms in its Trench, establishing an almost eerie similarity between the two countrysides, and I cannot quite shake a feeling that my choice of two watersheds was meant to be, be it by some pagan influence or by deities.

Maybe in the strange way of most unknown forces that become recognizable only when one studies them, there is even a third circumstance at work here. For lack of a better description we will call it the magnetism of Semiramis. For even more than the Vienna Gap, our flying leap brings us close to another reminder of our favorite goddess. It is in fact her home, namely the ancient and one-time notorious kingdom of Pergamon on the southern border of Celtic country fringing the Mediterranean. In words that could be an excerpt from romantic fiction, it was here that Semiramis was raised as a princess in a poor though royal family. Perhaps those early years explain her extraordinary drive for material achievement, so recognizable in this twentieth century.

But the Kingdom's own history describes its scramble for respect even better. In his book *The Celts,* Gerhard Helm relates Pergamon was the outcome of a shady transaction in a grand manner. Its founder was a eunuch named Philetairos, treasurer for a Celtic potentate whom he left when the benefactor's luck ran out. Helm writes:

> Any assumption that the Galatians (Celts) led a peaceful existence in their Anatolian (Mediterranean) refuge, with a wholly bucolic year interrupted only by ceremonies amid the rustle of oak-groves, misconceives

not only them but the time and place in which they lived. All around them flourished states founded on no more than armed force, or indeed on mere money. Are we to suppose that these Celtic warriors turned their curved lance-blades into ploughshares?

Although Semiramis had vanished from the scene long before the misdeeds of her family's bizarre kingdom had played themselves out, continuing events brought further examples of the evils to which she had been exposed. A principal performer was the eunuch's nephew, Eumenes, who used his late uncle's embezzled money to make himself the most powerful man in Asia Minor.

"Then he set about systematic extension of his possessions," Helm reports:

> He founded new cities and reinforced his army, still using the contents of the well-filled chest his uncle had left him. The embezzled money was also his best weapon against the Celts. Now settled in Galatia, (further east) they regarded Pergamon as a golden milch-cow, to be milked at will with the assistance of sword and lance. Not even Eumenes could deal with them: to keep them away he had to pay, pay, and pay again.

Occupying the region of Anatolia, south of the Black Sea and thought to be centered around Ankara (then Ankyra), a Celtic gang known as the "three tribes" was an early-day forerunner of organized crime. Here again we find the name of our dear friend, Semiramis, but very much in the past tense. Even had she been alive it is unlikely she would have had anything to do with an evil crowd as coarse as the three tribes. On the contrary, she would have hated them bitterly. They were invaders of her beloved land of Chaldea which dated back to the prehistory days of Akkad and Sumer at the confluence of the Tigris and Euphrates Rivers. There earth's first civilization had appeared more than 3,000 years before. Besides, she was an aristocrat on the highest level, and distinction between classes in her era was so great it was almost as if those of lower rank did not exist, except as the slaves which most of them were.

In the temple city where she lived with her lover, Attys, she was known by her other name of Cybele, or Agdistis as she was known to the people of Gaul who came from the north about the same time as the Celts. But even these newcomers regarded her with respect for their own religion spoke of "the great mother," which was the meaning of the name they gave her.

The position of her lover Attys was equally exalted. In fact it was his presence in the Pessinian temple, as the citadel was called, that more than any other consideration would have caused the Celts to be careful about what they were doing, if it did not discourage them completely. He was master of the temple which also was considered one of the most sacred monuments in the Roman empire. Archaeologists believe its relics of black stone are those that have been found in Ankara.

Referring back to *The Two Babylons* we find further reference to Cybele and Attes, with a note on the extraordinary blending of Christianity and Paganism arising from their relationship. To quote:

> Agdistis, that is Cybele... obtained from Jupiter, that no part of the body of Attes should either become putrid or waste away. Thus did Paganism apply to Attes 'the sinner' the incommunicable honor of Christ, who came to save his people from their sins'—as contained in the divine language uttered by the 'sweet psalmist of Israel,' a thousand years before the Christian era.

Thus, for a brief moment in our story, we have a glimpse of new enlightened times to come. But the moment of conversion is not at hand. In fact, far from it. Not only is the pagan belief still strong but so are even its most superficial trappings which are eagerly acquired by Julius Caesar himself when the kingdom of Pergamos is willed to the Roman people in B.C. 133. In *The Two Babylons* author Alexander Hislop says of Caesar at that time:

> Then, on certain occasions, in the exercise of his high pontifical office, he appeared of course in all the pomp of the Babylonian costume, as Belshazzar himself (last Babylonian king) might have done, wearing the mitre of Dagon and bearing the keys of Janus and Cybele (pagan adornments). Thus did matters continue, even under the so-called Christian emperors who, as a salve to their consciences, appointed a heathen as their substitute in the performance of the more directly idolatrous functions of the pontificate (that substitute, however, acting in their name and by their authority), until the reign of Gratian, who was the first that refused to be arrayed in the idolatrous pontifical attire. . . .

3

Enki and RJ

In setting a time frame for scenes featuring the ancient god Enki I have been guided by the "Time Chart" in *The Wars of Gods and Men* by Zecharia Sitchin. Without this Time Chart it would have been virtually impossible to give readily a useful dating of the periods concerned. I am grateful for the help given by this exceptional work of Mr. Sitchin.

In addition, it is his work throughout the three books of his invaluable series *The Earth Chronicles* that has enabled me to give historical substance to much of this chapter and the facility to carry occasional quotes. I am grateful for that help as well.

When I discussed the beautiful, artful Semiramis to introduce the most ancient phase of my ancestry I had little idea her story would remain alive beyond those early pages. But here she is again with the lingering fragrance and charm of a fresh flower pressed between the pages of an old book. It is not so much her beauty that brings her back, though that too is unforgettable, as her will to survive, and her cleverness. In other words, it is the same skill she showed in manipulating her beloved mysteries so that the world believed she had borne a white child to a dead husband black as coal.

While the star quality of Semiramis has, I think, amply justified my faith in her as a curtain-raiser for our ancestral story there are other reasons that would have supported my choice just as well. Foremost of these is that to an almost eerie degree she personified the family trait of liking to build and innovate (not a trait I inherited, by the way). Of all people, it seems a most unlikely part in the ultrafeminine personality of this bewitching woman. Yet an encyclopedic reference to her says:

> The name of Semiramis came to be applied to various monuments in Western Asia, the origin of which was forgotten or unknown. Ultimately every stupendous work of antiquity by the Euphrates or in Iran seems to have been ascribed to her—even the Behistun inscriptions of Darius.

It is a pleasure to note that her feminine allure still manages to show through, as the reference continues:

> The irresistible charms of Semiramis (which belong only to the legends), and other features of the legend, all bear out the view that she is primarily a form of Astarte, and so fittingly conceived as the great queen of Assyria."

Then comes more about our goddess as we expect her to be—and not to be. A column discovered in 1909 describes her as:

> ... a woman of the palace of Samsi-Adad, King of the World ... King of Assyria ... King of the Four Quarters of the World. ... The dedication of this column shows that Semiramis occupied a position of unique influence, lasting probably for more than one reign. She waged war against the Indo-Germanic Medes and the Chaldaeans

(presumably an error as she was Chaldaean herself). The legends probably have a Median origin.

Now, can the gifts of such stupendous talent reappear in a descendent arriving later by almost uncountable generations, and can those gifts be so special they are like a bequest from an ancient god? So unusual are certain factors in this case it is almost as if Enki left off where the mortal but multitalented Semiramis began (though she was a goddess to Chaldeans). Also it is as if my father had some of the benefit of their largesse, though he was of course not related at all to the immortal Enki and only distantly to Semiramis by virtue of her adopting the name Magor.

Yet a likeness to Enki in a minor key is unmistakable. To underline that I make reference to some of the epic work of Zecharia Sitchin whose masterly study of ancient gods seems to bring these deities to life as if they are touched by the same magic they bestowed on others. So comparisons with mortals become feasible.

In *The 12th Planet*, which gives readers a spellbinding account of prehistory air travelers arriving on this planet under Enki, Lord-to-be of Earth, Sitchin quotes him:

> When I approached Earth,
> there was much flooding.
> When I approached its green meadows,
> heaps and mounds were piled up
> at my command.
> I built my house in a pure place. . . .
> My house—its shade stretches over the Snake
>   Marsh. . . .
> The carp fish wave their tails in it
> among the small gizi reeds.

Because Enki so favored Earth's coastal areas, Sitchin explains, he was also known as "lord of the watery deep." The marshland, he is quoted as saying, "is my favorite spot: it stretches out its arms to me." He then tells how his crewmen "drew on their oars in unison," how they would "sing sweet songs, causing the river to rejoice." At such times, he said, "sacred songs and spells filled my watery deep." He named his boat MAGUR, meaning a "boat to turn about in."

It is now more than 445,000 years since Enki landed on Earth—to be specific it is a summer evening in the early 1920s—and we see Robert James Magor, or RJ as he was often called, preparing to leave his office in Montreal by putting into his briefcase a folder of photographs he had just received from the printers. Already a successful young businessman, with a large house on the fashionable upper slope of Peel Street, he had married a beautiful American girl, Marion Ferguson, who by then had brought six of us into the world. Soon she would have a family of seven without losing her stunning good looks.

The photos he was taking home marked another upward step in business for still youthful RJ. They were a series of shots showing a trim outboard motorboat that was the start of a new manufacturing line at his plant in Hamilton, Ontario. Until then the plant had operated exclusively for production of the railway cars that gave his company its name of National Steel Car Co. Ltd., an affiliate of the Magor Car Corporation already operating in New Jersey. In short, he was a builder and innovator with an awakening interest in boats.

Although the manufacture of boats would be barely more than a sideline, it seemed almost as if the gods had ordained it as part of our father's destiny. There in the heart of Canada, hundreds of miles from the sort of coastal waters so favored by his "ancestor" Enki, he was undertaking to build a craft that surely would have gladdened Enki's heart. It would be a flat-bottomed lightweight skiff that could navigate through the marshy waters of Laurentian mountain lakes for which it was primarily designed. Its first testing would be in just such a body of water—Lac Echo in the Laurentian Mountains where our family had a summer cottage. While my father was not a fisherman and had no obvious reason to prefer Lac Echo when his friends and other Montreal businessmen found their summer enjoyment on more distant ocean beaches, he remained loyal to that quiet little lake. Occasionally, as a family, we were taken at first to more popular summer resorts, but Lac Echo eventually prevailed. So much so that a complex of Magor, or related, cottages developed there. Our father ran things at one side of the lake and his sister, Sophie Dodwell, ran things at the other. Oldest in a family of fourteen children, Sophie was a born leader like our father, who was the youngest. Her personality was so dominant that she, like him, gave a Magor identity to most communal affairs at Lac Echo.

Strong as his family ties were, however, I suspect RJ's ancestral instincts were at least as strong in bringing us back to that spot summer after summer.

First there were those boats he was about to produce. The trade name he chose for them was "Sea Duck." Considering our distance from any body of salt water, it was an unexpected choice, but it was much less so if we think of Enki in his MAGUR. The "lord of the watery deep" could hardly have improved on "Sea Duck" to describe a craft made for bobbing about in coastal areas much as our boat would bob about near the shores of a lake.

Significant, too, was the fascination Dad felt for the ancient Laurentian Mountains. I remember our geology instructor at Columbia University telling us that nowhere on the North American continent was there a prominence on which age had a greater part in deciding its shape. The rounded slopes are now just mementos of a time when ice came to chisel new features on the face of the continent. That was so long ago it is difficult to understand that Enki first saw Earth after an ice age had come and gone. That would be followed by another glacial period and then a third!

Coming down in the Arabian Sea in the wake of the first freezeup, Earth's pioneer immigrants looked out on a chill and watery land that would need human hands to make it habitable. It was in that improbable setting that Enki went on to build cities. He called the first of these Eridu, meaning "house in faraway built." It became Earth Station 1, the center of this planet's aerial operations.

Before RJ took charge of our place in the Laurentians, the setting there might well have been a scene similar on a minor scale to what Enki faced at first. Reeds were always present and, in place of a snake marsh, we had a leech marsh extending from the lake. Most of us would avoid swimming, if on any given day, the "bloodsuckers" were particularly bad. In the center of the compound, RJ arranged for a landfill of gravel so a tennis court could be built, and all around stood the Laurentians, lovely now but relics all the same of a once frigid, hostile climate.

Strangely, however, it is right at this point where a quiet country life seems to be taking over that we can again pick up a comparison with the vivid ancient image

RJ as "Enki" guides the canoe to safety. 1938 Lac Echo.
—Felicia Magor

of Enki. Just as Enki's life was a study in contrasts between action in dangerous places and repose where little stirred, so was RJ's. For he was an outdoors man by instinct, and it was not on the placid waters of Lac Echo that this feeling was stimulated, but on the wild waterways of northern Quebec which, as a young man, he had seemed driven to explore.

His boat for that purpose was one of Indian design and once, in an unusually frank moment, he told us how he had navigated rapids and even an occasional waterfall in his fragile craft.

Our sister, June, recalled it was a matter of record he was the first white man to guide a canoe through the tortuous Lachine Rapids on the St. Lawrence River outside Montreal. His skill in waterborne craft was further recognized when he was made "stroke" for an award winning eight-man shell at the Lachine Rowing Club.

To an uncanny extent it was as if at venturesome times in his past that Magor was acting out a contemporary version of what his predecessors, particularly Enki, might have done in an era before recorded history. Yet his most eerie tuning-in to those distant ages was still to take place. Twenty years or so from the days at Lac Echo war would explode around the world and Magor, as if at the special urging of Enki, who was a warrior much more than he was a devotee of placid waters, would enter yet another field of building enterprise. He would build aircraft and he would choose "Lysander" as the name for his first creation. Somehow it appealed to him to turn to mythology for a way to identify this major undertaking, just as he had taken a name from distant water to identify his boat.

With the outbreak of World War I, I too, experienced a turn of events that might have pleased Enki, if he had known anything about it. For in two widely separate postings as a pilot in the RCAF I was assigned to flying boat squadrons, the first operating off the Canadian west coast and the second over the North Sea followed by operations over the North Atlantic. If I had known about Enki in those years I would have thought of him particularly over the North Sea. For it was there our four-engine aircraft took us into gales that blew with such special fury the sea would lie flat under their anger and for a while all would seem peaceful. Perhaps that was Enki's way, though I do not think of him now as a deceitful god. Maybe it was simply a matter of not showing his true feelings, as I think was the case with RJ.

I reflect in this manner because after I had immersed myself in a study of gods, I began to believe that perhaps they had always been closer than I realized. That is not to suggest I had been singled out in any particular way but it could mean that an influence is always present for most of us which may require only a slight acknowledgement to give it meaning.

For instance, in earlier days of flying I thought it was merely chance that I was posted to a training school at Malton, Ontario. The place then was just a few farm buildings on an open field. Even when they called it an airfield it was so unimportant there were no runways. In winter, which seemed to be most of the time, we would rattle about on crusty snow in planes equipped with skis, and it was a morning ritual for the chief instructor to take off first to make sure everything was all right for the rest of us.

That was the Malton that soon became a major air terminal and simultaneously the center of RJ's aircraft-building enterprise. Although there were other places where I might have had my elementary flying training, the family gods or caretakers chose to stage this coincidence at Malton. It meant nothing to me until I realized from that and a myriad of other small experiences that something odd was going on. Once it had sunk in I was content to go along with the plot, whatever it was. It would have been a mistake to believe out of hand that no possible benefit would come from doing so.

What influence except their own, for instance, led me to learn about Enki, and others? Had I protested against their subtle designs, I was sure my ancestors would find a way to say "go your own way, then," and I would not have liked that. Had I not been guided, I may never have known these ancestors, not even their names.

Still, the question can be asked, what was the purpose of that relationship? I, of course, did not know nor would anyone else when faced with a question that so concerned their own destiny. The more we study ourselves, the farther from the answers we stray because we won't tarry long with anything that is upsetting.

Instead, I thought, most of the answers seemed to lie unattended in the soft matrix of our sensibilities and in the more unyielding one of ambition. Their formation may have started at the time of conception but I suspected that with each person they were the end product of a long ancestral line, tangled in some places but for the most part usually intact.

To act contrary to the messages it transmitted would surely lead to discomfort at the least, or to disorientation in various forms as the more likely result.

If I was ever doubtful of the part that the spirit of ancestors had in my family's affairs—and I am sure that on reflection many families could bring to mind a similar presence in their own lives as well—I needed only to think of RJ. That was particularly during the little time left to him after his decision to join the events of war. It was a time with a strong mix of tragedy and fulfillment emerged to form the climax of an outstanding life. What developed might well have been an episode from Sitchin's book *The Wars of Gods and Men*.

As demands of conflict mounted and more help was needed from overseas, RJ's manufacturing skills were called on to produce a four-engine bomber in addition to the Lysander which was already in combat. Under tremendous pressure he organized his new plant at Malton, to produce the heavyweight Lancaster bomber which, in a miracle of factory speed, was in time to take a full part in the quickening pace of Allied operations over Europe. It was a project that in energy and vision may have equaled what Enki needed in prehistory after landing on Earth, though Enki's purpose then was to build cities, not to prevent others from destroying them.

(But the destruction of Eridu, Enki's first Earth station, by mortals did occur. Sitchin's *Earth Chronicles* record its civilization perished under a "radio active cloud" in 2023 B.C.)

So great, however, was the pressure on RJ to meet production deadlines that he succumbed to the strain and died in hospital at Montreal before the Lancaster helped achieve the Allies' victory.

Then Mother saw to it that his last wish was granted, which was that after cremation his ashes be buried at Lac Echo. For a marker she chose a stand of silver birch overlooking the water and all that was done among the ancient mountains that may have brought to RJ an ancestral recall of time before the ice ages.

Later I was discussing that period with my brother Lincoln who had entered the industrial scene in Toronto on returning from service overseas in the Canadian Army.

"It was a difficult time at Malton," he remembered, "keeping up the pace with a war on and everything in short supply. Then the British group under A. V. Roe—or Avro Canada, as they called it—took over and things were a little easier with a whole company doing the work of one man.

"As you know, the first president was Crawford Gordon, and it was he who promised a flying machine so revolutionary there would be nothing in the air to match it. He said it would be a 'flying saucer.'

Thus were the gods emulated—or at least an attempt was made—with none other than Field Marshal Montgomery being among the first guests invited to watch the launching of a "Canadian flying saucer" capable of traveling 1,500 miles an hour and flying straight up. Viewing plans were so tight that two of Montgomery's escorts were barred from the spectators' area at Malton. But the excitement ended right there anyway.

"As far as anyone knows, the Avro saucer never left the ground," Lincoln explained. "But, like others, I agree that a working model is possible. I'd like to give it a try myself." As it turned out, it was incorrect to say it never left the ground. In a film shown in 1991 by author Pierre Burton on the history of Canadian aviation, an experimental "saucer" is seen briefly, wobbling to a height of about four feet.

Lincoln Magor

Did I hear a bit of RJ in that remark? Of all the family, Lincoln seemed to have inherited the most obvious strain of his father's manufacturing ability as even then, still in his thirties, he was president of a highly automated engineering plant. Its equipment could be made to cut a small-scale replica of almost any piece of machinery fed into it, and for that it had the clever name of MIMIK.

Perhaps now that his attention had been caught, Lincoln was being told by ancestral reminders that followed that the time would come when he could use that name in a strikingly appropriate way, which would be for small aircraft capable of big performance. Although the pressure of other business would put the matter on hold for some considerable time, it was curious how the reminders persisted and were presented in an imaginative way as if to make sure they were remembered.

There were two after this one, both occurring as UFO incidents. Although neither had an overt association with Lincoln, instinct told me he should know about them. So I took care to show him my write-ups like this one as I

Lincoln Magor

believe I was meant to do, for much of the caretakers' work is done by accomplices, consciously or otherwise.

The first reminder came as part of the great UFO flap of 1967-68, a period of so many sightings that every interested observer knew something special was happening, particularly in Canada. It was as if the aliens, using some kind of rotation system, chose this country for attention because they had never done so before. In *Canadian UFO Report* we called it "The Year We Were Invaded Without Even Knowing It," Vol. 1, No. 6.

However, a number of witnesses knew it quite emphatically, and one in particular was Joanne Hammond, a young woman of Radium, B.C., who was driving alone one night in this period. By curious chance, and for her it was poor luck, she was on her way to visit friends in a mountainous section we had already named "Valley of Gods" for the very reason it was occasionally the scene of unexplained aerial activity.

"Something with wings suddenly flew head-on at the car," she told two of us who talked to her later for CUFOR. "When it first came toward me it looked round with a hump on top. But when it shot over the car it looked more triangular. The two wings slanted backward and right behind in the middle was a narrow tail about six inches long. The wings spread across the windshield, so I guess the whole thing was about four feet wide altogether and it had a light coming from the top of it, so I knew it wasn't a bird. But even though my window was open a crack at the top, I didn't hear a sound, yet it was just inches away when it dove at me."

Joanne Hammond

So I knew, too, we were probably dealing with a true case of a UFO encounter.

After almost smashing into Joanne's windshield, the thing shot up and disappeared momentarily.

"But I knew it hadn't gone because I saw the light in the rearview mirror," Joanne added. "It lit up the whole inside of the car. I was doing about seventy miles an hour by this time and still it came after me."

After more than two terrifying miles to an intersection where houses came into view the object gave up the chase. However, the experience so frightened Joanne that for months she would not drive by herself at night.

Joanne Hammond

The second incident involving small aircraft that I put on record, particularly for Lincoln's benefit, had considerably more emphasis on detail. In fact, it was a display piece for anyone who, as in his case, might have been contemplating construction of such craft. Moreover, the incident had the benefit of an expert's observation supported by that of other witnesses, so there was no way to dismiss it as UFO daydreaming.

Place of the sighting that occurred on a clear summer evening in June, 1976 was a large car-sales lot in Victoria, B.C. About fifteen bystanders were present among whom was one Cliff Bennett, holder of a private pilot's license and a member of the Victoria Flying Club. Bennett gave a report for the group to the Canadian Forces Base at Esquimalt. Released later as a CFB statement, with details of appearance strikingly similar to what Joanne saw, it said in part:

> They (two objects) appeared to be the size of large birds with swept-back gull-wing configuration. There ap-

peared to be virtually no fuselage, wings being the predominant feature. The underside appeared to be lined with neat rows of vents. Observer claims he noted that leading edges were thick and rounded, tapering to knife edges at trailing edges. Observer was impressed by the fact that despite the rapid and random nature of the maneuvers executed by these objects, there was absolutely no flexing of the relatively thin wings. There was no noise of any kind accompanying the maneuvers.

A second type of object appeared ten minutes later. It was a bluish-white sphere with a white corona. Observer felt it was "somehow in control of other objects sighted." It remained stationary for 45 minutes and then moved away parallel to the earth. Meanwhile one of the small objects disappeared in another direction. Sight of the others was also lost as they were so thin that when they presented a fore and after line to the observers they could not be easily seen. Duration of the sighting was just over an hour.

All that was more than ten years ago and, without any more strange incidents to record, I had almost closed the book on the subject of family caretakers when suddenly they were back, or so it seemed. This time it looked as if they had made a surprising approach through our sister June, who had not been involved earlier in their mission. A sign they wanted action may have been the significance of a *New York Times* clipping from her to Lincoln that said the U.S. Navy had started to use robot aircraft in a theatre of operations. Specifically, that was aboard the battleship Iowa in the Arabian Sea.

Had the caretakers egged June into sending an apparently impersonal item as another reminder for Lincoln? An overture through our sister, who always showed particular concern anyway for her family's interests, would be a tactful way for them to turn matters to their favor. Although they usually preferred a dramatic impact for their reminders, as we have seen, they were also capable of a more subtle strategy and apparently had used it effectively for most of RJ's life.

So it was not really surprising when we learned soon after that Lincoln had reorganized his business to go a step further by producing a revolutionary new line of vertical flying machines. To start things off a prototype engine was fired up on New Year's Day, 1988. Obviously, I thought, Enki had something to do with the venture.

For just as RJ was moved by an ancestral spirit and selected names from distant places, like Sea Duck, from an ocean that wasn't there, and Lysander, so too did Lincoln. For his new aircraft with its origin of mystery he chose the name Merlin!

Knowing the caretakers' cunning ways by now, I felt almost as if they were reading over my shoulder when I came across a magazine article about "incandescent followers" chasing aircraft in World War II. Even more to the point, as I was sure my unseen companions would realize, was prominent reference to a Lancaster bomber almost burning out an engine while shaking off one of these "foo fighters" as the phantoms were called.

Had the pilot only known, I thought, the ghost was probably friendly. Maybe it had decided, perhaps correctly, the plane was one of the Lancasters built by RJ before he died and it was providing escort!

Written by Andy Roberts, editor of England's *UFO Brigantia,* and published by the *Mutual UFO Network* in its Journal, April 1988, the article said in part:

> Balls of light would appear from nowhere and play tag with aircraft for up to forty minutes. They were not hallucinations, being in some cases seen by the entire

crew of a Lancaster bomber, and were not reflections as they were seen from different angles or from two planes at once.

Evasive action to shake them off was no use. In one case a Lancaster almost burnt its engines out, going "through the gate," a slang term used by pilots to denote pushing the engine to its limits, in an effort to lose its incandescent follower, but to no avail. None of my respondents (aircrews questioned) had fired on the phenomena, in some cases fearing it to be a secret weapon which would explode when fired upon and in others just attempting to evade it on the basis that as long as it wasn't firing at them they weren't going to antagonize it. . . .

I firmly believe that foo fighters were a real, although nonsolid phenomenon and I reject that hallucination/misperception hypothesis almost entirely. These people's lives depended on being able to see and identify aerial objects very quickly. One mistake and it was their last.

As I drew on the charms of Semiramis to open this chapter, it seems fair to also close with her. To do that I note the curious fact that fish had a place in her life, as well as in the lives of Enki and RJ. We know by the perceptive work of Zecharia Sitchin that Enki observed fish in coastal waters almost at the moment he landed on Earth, and RJ's Echo Lake abounded in fish.

Now I see that in the case of Semiramis the importance of fish surpassed both of theirs by far. For she was the daughter of the fish goddess Atargatis. That association is noted forever by a statue in the great temple at Hierapolis. To top things off, I as your humble narrator was born under the sign of Pisces, the fish.

4

A Journey into Myth and Nationhood

Although Lincoln tends to discount the significance of family gods in our lives what other explanation can there be for the almost magical surge of good fortune in his own affairs? It came with his development of a little machine-tool just at a point when his other business affairs looked gloomy. As Lincoln takes up the story:

Dad died in 1942 when brother Robert (now deceased) and I were working the aircraft division of National Steel Car in Hamilton. The division was spanking new and owed its existence to Dad's foresight and energy. I then went overseas with the Royal Canadian Artillery and served as the Technical Adjutant of a regiment of self-propelled anti-tank guns.

On my return to Canada in 1945, I went to work in the machine shop at National Steel Car in Hamilton and subsequently was posted to Vancouver as the company's lumber buyer. If Dad's successor could have found a Canadian city even further away, I'm sure that's where he would have sent me!

The high grade lumber required for box-cars was in short supply on the B.C. coast so I explored the interior for alternate sources. I found what I wanted but it was difficult to get at and roads there were rugged and costly. I considered the idea of flying the lumber mill into the site by helicopter and then taking the dried and dressed lumber to the rail head the same way. The helicopter industry was then in its infancy but its future, as has been proven, was very bright. The aircraft division of NSC had just been sold and the company was sitting on a pile of cash which I thought management might consider investing in a growth situation, but I was wrong.

I left NSC and, with the help of a few friends, started Magor Aviation Ltd. to design and build a prototype of a new German concept in vertical lift aircraft in which the blades rotated around a horizontal axis rather than the vertical one of conventional helicopters. The Canadians at the National Research Council in Ottawa were convinced it wouldn't fly, but the Americans at the U.S. Naval Research Center thought it would. Unfortunately, it never made it past the budget committee.

To keep my little company afloat, I brought in subcontract machine work from the aircraft industry and in the process developed a machine tool product, called MIMIK, which allowed a machine tool to reproduce parts for a model or master. In subsequent years we sold many millions of dollars of MIMIKS all over the world and right now I am working on a version which can be interfaced with a computer. That's the patent work I told John about. Some of the more interesting applications of the MIMIK are—reproducing, in stainless steel, the femur of a cancer victim, machining the base ring of the Apollo moon rocket, and, hopefully, the removal of the weld bead inside the space shuttle booster rockets.

Had Dad lived I'm sure that, with his imagination and drive, National Steel Car would now be in the aerospace business in a big way and the family tradition of exploring uncharted paths would have carried on.

But though it was not uncharted paths they placed before us it was the fascinating paths of ancient history. (Here John continues with his story.) Those family gods are indeed on hand and listening, exactly as I thought. What other explanation can there be when the opening line of Kate Seredy's wonderful book which I was just starting to read said, "Those who want to hear the voice of pagan gods in winds and thunder can follow the thread on the pages of this book?"

I picked up the thread, and how could I resist? My sister Felicia and I were in Budapest at the time, having come to stay in our ancestral homeland for a while and the book was *The White Stag* by Kate Seredy. As I've said, one of those in the book who listened to the pagan gods was named Magor. It was he who was to lead followers, known by their tribal name of Magyar, from the snows of Asia in search of a more pleasant land far across mountains to the west. With him would be his brother Hunor whose followers were called Huns. Lasting for many years, their long march, which would be completed by their kin of later generations, would end in their finding a new home on the shores of the Danube.

While I had, of course, heard the story before, I had never heard the fascinating way in which it could be told, and here I mean the flow of Seredy's writing. Also there is the enticing promise of mysteries beyond. In her hands, the White Stag quickly becomes an elusive form that must be followed, and I find the effect an intriguing one. When the brothers return from a pathfinding trip, Hunor reports what they saw:

Seven moons ago a miraculous White Stag appeared on the crest of the hill. He was white as driven snow and bigger than any stag ever seen by man....

All night he ran through the forests and plains across rivers and over mountains, and we rode after him as we had never ridden before. The hoofs of our horses never touched the ground, we soared over valleys and left mountains far below us. When morning came the White Stag stopped on the edge of a misty blue lake. As he stopped, our horses fell back exhausted. They stumbled and snorted and would not move again. The White Stag pawed the ground where he stood and shook his antlers, then he disappeared in the floating mist over the water.

All that day we searched for him. We did not see him again. We only found the place where he had pawed the ground. There were seven deep rifts in the ground, cut deep and wide as no living beast could cut them.... There were trees heavy with fruit and the air was sweet with the breath of beautiful flowers.

Thus, far ahead of any others, Hunor and Magor had found the land that some day would become Hungary. In the deep cuts of the earth supposedly made by the White Stag, they had also found the traces of a mystery or a mystery described in a way that was new to me. "There were seven deep rifts in the ground" Hunor had said, "cut deep and wide as no living beast could cut them."

As the reference to "no living beast" made me curious, I mentioned it in the course of a letter to a friend, Kazmer Ujvarosy in San Francisco. He not only is a Hungarian-born scholar of wide learning and a linguist but also has an interest which I considered of first importance. Like me, he is a student of the UFO mystery.

With my letter I enclosed a few lines from my proposed book which contained a reference to the early-day

Magor and the White Stag story, my hope being he would comment in any way he chose. I was not disappointed, as this was his reply:

You mention a "White Stag" that led Hunor and Magor to a new land, just as a mysterious object led the wandering Israelites in the wilderness. To be precise, the legends speak of a *Csodaszarvas,* "Miraculous Stag" or "Marvelous Stag." Obviously the stag was miraculous or marvelous because it was not a stag. For example in the *Coptic Texts on Saint Theodore the General,* translated by E. O. Winstedt, the stag has the characteristics of a UFO. When St. Theodore and his 500 soldiers lost their way in the desert, and the Saint went to look for help, a voice came to him from heaven, saying:

"Theodore, son of John the Egyptian, cease from advancing in this desert with thy soldiers, for the tempter is tempting thee." And the Saint hearing this was astonished, especially at hearing the name of his father, and he said: "Who is it who called the name of my father?" Straightway, lo, a stag appeared to him like the orb of the sun when it rises on its basis. And when St. Theodore saw the young stag in this form he was astonished, and again he looked at the shining car that was above the horns of the stag. Straightway the lamb that was in the midst of God who taketh away the sin of the whole world. I am Adonai, the lord Sabaoth, the God of John thy father."

When St. Theodore heard this he turned his horse and went back in fear to his band of soldiers. He found them and their beasts lying like the dead through their thirst. And St. Theodore got down from his horse and offered a prayer saying: "Lord God, who raised up those that fell on the way to Babylon, and gave thy people strength in the desert forty years without suffering, raise

Kazmer Ujvarosy

up for me these who have fallen through my undertaking. Thine is the glory for ever, amen."

Straightway, behold, a cloud of light shed dew over them, so that their hearts were strengthened like men who have drunk their fill of water cool and sweet. And he (the Saint) said to them: "My brethren, it is the gift of God, that has raised you. Now come and see this beast, which I saw on the top of the mountain. I never saw one like it: most wonderful was the young stag which I saw." Straightway he went with them and took them and shewed them the stag he had seen. And they marvelled saying: "We have never seen one like unto this in its graceful form." Nor did St. Theodore know this meaning of the lamb which spoke with him. Then he said to his band of soldiers: "Let half of us get behind this beast and guard it that it escape not: and look ye, strike it not with warlike weapons." Then the army formed two companies and left the stag in their midst, if haply they might catch it. And the Saint raised his eyes aloft to the sky and prayed to the Lord for the young stag. Straightway he saw the Lord Christ in the form in which he had seen him above the stag. . . . Again he heard a voice in heaven saying: "Theodore, Theodore, my beloved, . . . I am the stag which thou sawest upon the top of the rock. Now be thou valiant and suffer martyrdom for my holy name, and behold I will grant thee and thy comrade the Eastern the grace of my great Archangel Michael that your souls shall be on his right hand in heaven: and every war into which ye enter to fight, I will send the Archangel Michael to crush and scatter the armies before ye, till your name is famous over the face of the whole earth to all generations. Because thou hast believed in me, I will save thee; thou has called unto me and I will hear thee. Now, my chosen Theodore, behold the beauty of this stag and what like he is in his form." And St. Theodore looked towards the stag and saw the fiery car above its horns. The Lord said to him: " . . . Return to thy band of soldiers and tell them to cease from pursuing after this stag." So he went to his army and said to them: "My brethren, trouble yourselves no more in pursuing after this stag: for we shall not take it, but it will take us in the nets of its goodness." And he told them all that he had seen and heard from the Lord. And St. Theodore came down from the mountain with his soldiers and let the stag go.

According to a Magyar legend Prince Geza of Hungary and his brother, St. Laszlo, also had an encounter with a peculiar stag. When they were discussing a war plan at a strategic location just outside the city of Vac, the Lord made known to them in a vision that Geza was destined to receive the crown of Hungary. In return Geza made a pledge to build a cathedral for the blessed virgin mother of Christ. Following his coronation in 1074, King Geza and St. Laszlo returned to that place where the heavenly vision took place, and they were pondering where the cathedral should stand.

Standing there, all of a sudden a stag made its appearance before them, whose horns were full with burning candles; as he began to run away from them towards the forest, he stopped short on that very spot where the present monastery stands; when the knights shot arrows at the stag, he leapt into the Danube, and they saw him no more. Seeing this, Saint Laszlo said: "To be sure, this was not a stag, but God's angel." In reply King Geza asked: "What could all those burning candles be, which we saw on the horns, but glittering feathers, and where he stopped, there he marked the place, that we build the cathedral for the Blessed Virgin there, and at no other place!"

As these stories indicate the *Csodaszarvas* that Hunor and Magor chased was not a stag, but a brilliant flying object or "fiery car." Evidently it resembled a stag because it had curved wings (horns) and its movement resembled the wavy motion of a running stag. The claim that the Lord Jesus was in the fiery or shining car suggests that Christ played an intimate role in the destiny of the Magyar nation.

When I was publisher of *Canadian UFO Report* it was always a pleasure to ask Kazmer Ujvarosy for an article touching on ufology or some related topic. With a fascinating book on the life of Christ to his credit, he is a true scholar and an original thinker, and my request never failed to bring a rewarding reply.

So I made it a point to have something by him in this book. The contribution here is only a part of what he sent and there will, I hope, be an early occasion to carry the rest.

But being well pleased by what I had already received and continuing to have reference to Sitchin's work for the events of prehistory, plus having now read Seredy's book, my curiosity about the White Stag was aroused to a point that I imagined a scenario on the subject, and this was it:

It is a night 400,000 years ago in the high lands of what is now southwest Asia. The time of the Deluge has passed, yet a bitter wind still sweeps at the tattered remains of the dark cloud that came to drench the world. The ground is hard.

As a few shivering human figures huddle in a cave they are frightened by a blazing white light probing the gloom. Without a sound it travels above them and moves towards a place where the ground is flat. Now the shivering humans can tell the light shines from a massive object so gleaming white it almost shines through the darkness.

The white object comes closer to the ground and finally touches it, all the while in silence. The humans see the giant bird—and they have no other way to describe it, though it has no wings—shudder and sink a little into the hard earth. They know by the cracking noise of the earth that this is a heavy bird indeed.

After a while a place in the bird opens and humans emerge. To the watchers' amazement they float to the ground. At last when there are no more of these another larger human appears at the opening. He has a white beard and a voice louder than the others as he seems to tell them what to do.

As the watchers will one day learn, this is Enki, a god from another world where he is called Lord of the Saltwaters. Still later, long after the strangers have gone, the watchers will try to draw on flat rocks and on the walls of their caves what they have seen. But how do they show a bird so large, and how do they know it was a bird since they had never seen one like it?

As it had no wings they decide instead it was an animal, an animal that could fly. For that they draw a leaping stag and their stories of what they saw last forever.

I pinpointed southwest Asia for the scene because Enki's craft was thought to have landed in the Mediterranean area (*The Wars of Gods and Men* by Sitchin) and Seredy writes of the lasting tracks it left, which could mean the ground was frozen. Southwest Asia would meet both conditions.

While details of strangeness like these in my imagined scenario were not taken from any one case on record, we should understand that in ancient times or in UFO incidents they may not have been strange at all. In antiquity, researched and described by Sitchin even the nefilim or lesser gods were capable of flight, and in authenticated UFO cases the silence is commonplace.

There is another speculative detail that I am tempted to include here. It concerns a wise and respected man in Enki's time named Adapa who was chosen as Enki's second-in-command. It also concerns a wise and respected ruler of Hungary whose name was Arpad and who was responsible for converting his once heathen country to Christianity. With names and qualities so similar, is it possible that the two great men were related in some way beyond our understanding across an enormous distance of time? Did Arpad receive his god-like legacy from Adapa? The curious possibility of greatness being transferred by rebirth, perhaps with the help of gods, has been noted in these pages before.

Making Arpad's case stranger yet, there was another touch of the gods in his background. In *1,000 Years of Hungary,* Emil Lengyl relates that Arpad's grandmother dreamed her son would sire a line of great enlightened rulers. Arpad, then unborn, was first in that line.

Before then, however, it was just mortals with mortals' ordeals who took the brunt of shaping an identity for their nameless country, and the man in charge was hardy old Bendeguz, who we almost forget was Hunor's son. Bendeguz held the loosely formed group together while stragglers picked up along the way swelled the ranks, as did others who were taken captive in hostile encounters. Yet all had one thought in mind. It was some contemporary version of "head for the pass."

But in that unlikely throng there was a happening that would be remembered through the ages. It was the birth of a son for a young Cimmerian woman in the captive group, who had won the hearts of all, and the father was Bendeguz. They named the child Attila and death was there to welcome the boy as his mother died when he was born..

In *The White Stag* Kate Seredy writes:

> On a summer night in the year 408 a flaming red comet appeared over Europe striking terror into the hearts of all who saw it; a menacing omen, a flaming red comet shaped like a tremendous eagle with a sword in its talons.
>
> In that year, when the walls of Rome were cracking before the onslaught of the Goths led by King Alaric;

when the Vandals were invading Hispania led by King Gunderic; when Roman Britain was fighting a losing war against terrible barbarian pirates, the Saxons—on a summer night of that year Attila was born. . . .

And at that hour, Flavius Honorius, the Roman Emperor, gazed out of the window of his palace in Milan long and fearfully at the flaming red comet. He knew that the great structure of the Roman Empire was trembling and cracking under his feet . . . might this fearful omen herald the end?

And in a dark tent, between the river Rha (Volga) and the river Tanais (Don), a new born child cried bitterly, cried for comfort and warmth and tender love, cried for the things he was never to know.

As the plodding long march of Bendeguz and his camp did ultimately lead to a home for Hungary finally won in battle by Attila, Hungarian scholars and explorers have tried for years to find their pioneering trail. So far all have failed, but there was one who may have come very close before death overtook him in the Himalayas where he was buried in 1842.

His name was Sandor Korosi Csoma, and a highly respected name it is in the world of science, Emil Lengyl writes. He was a scholarly explorer and a remarkable linguist who set out to seek the ancient home of the Magyars (as all the migrants became known in history) wherever the tracks might lead him. In the writings of medieval Arab scholars he detected references to the esoteric Magyar race. His contorted tours of exploration led him first to the Arab lands which today we call the Middle East, in search of authentic documents. Thence he penetrated into Central Asia, into the land of the steppes and of the valleys in front of the high mountains. The chroniclers spoke about the Uigurs, a word that sounded like Ungar, Hongrie, Hungary; and there he was looking for his kinsmen's forebears. Then he sought Hungarian kinship in Dzungaria, the remote land that lies northeast of Tibet's Lhasa, whose name may have been the origin of Hungaria.

Creation of the White Stag for the dramatized part of this Hungarian epic is a way of infusing continuous vigor into the story of a journey that takes us over the entire expanse of Asia and well beyond into Europe. It is an odyssey in which we cross two formidable mountain ranges—the Urals and the Carpathians—and see the span of a lifetime for the leading performers.

Nimrod, who was an old but vigorous patriarch when the long march began, dies at home before it is well underway. The aging of his sons, Magor and Hunor, becomes indefinite but Bendeguz, who was born en route, is an elderly coleader at the end. When Attila was a baby in that dark tent by the Danube, the march had not yet found a way through the Carpathian Mountains.

To gain an introduction to Hungarians and form impressions, Felicia and I stayed for a while in Budapest and decided we could spend years among these people without ever being able to say we knew them well.

We had an example of that in the first days of our visit. As our hotel overlooked a large city square surrounding a park we could quickly see what was happening below, which was never anything much. By day a few strollers would cross the park and by evening all was still. Except for one time when a strange discordant noise rose from the park and we saw a small crowd gather in front of a church. Felicia who speaks better Hungarian than I, though she doesn't speak it well either, went down to investigate.

On return, she said that after watching for a while near the church—it gave significance to the occasion as there are so few churches in the city—she asked a woman next to her what it was all about. Although the woman didn't speak English and the racket of the instruments made things worse, Felicia gathered it had something to do with the millennial celebration of Budapest's founding which was then underway.

But the woman sounded apologetic. "We do not celebrate very often," she explained, "so they do not play very well."

But as the world knows, Hungarians' appreciation of good music is something else, and so is their performance of it, which explains why we were in Budapest at the time of the Spring music festival. As the brochure said, the "festival brings the finest Hungarian and international performers to Budapest."

But in a typically Hungarian way, that was an understatement. What the brochure did not say, except in serving as a directory, was that performances were given at various galleries and concert halls of the older type, allowing the visitors pleasant exposure to life at night in that charming city as it was in the past.

Although evening dress in Budapest is no more formal than it is in most cities today, we felt we had taken a step back in time as we sat somewhat primly in one of the narrow tiers of white and gold balconies that seemingly had changed little from what they used to be. And we heard music that came with the thrilling fullness of Hungarian artistry.

That night it happened to be selections from *West Side Story*, a favorite in Budapest, and we were transported to some place between as the music of the west joined the mystery of the east. Much more than legend was given that evening to our ancestral image of the White Stag.

Long before that moving juncture, however, and in the confusion of changed faces all around me, I lost the guiding thread that Seredy's book had offered. The loss could have seemed serious and perhaps beyond replacing until I was given a substitute that might have done just as well. A friend suggested I call my quest a search for psychic roots, which was apt. It meant that while looking for ancestors I might come across a stranger who would fit into the ancestral picture far more securely than the "real thing." Enki was one of those, and so was Arpad of Hungary. Eventually, however, I settled for the book's present title on realizing it was not accurate to imply Felicia and I were making a search for something we did not have or did not want. Kate Seredy's single powerful word, resonant with distant chords, said it so much better. We were pagans.

5

Bell Jars

While the happiest marriages are made in heaven, as they say, there may be confusion at times among heavenly authorities about who is best for whom just as there is among mortals. Also the same confusion might have existed for pagans, who listened to different authorities.

Thus there was almost certainly a place for marriage counsellors, probably better known as religious advisors or something similar, even in those distant days. For instance we have seen how Semiramis and Bendeguz seemed to be made for each other, and it would be natural in such cases if the advice of elders was sought.

But there was another man with whom Semiramis might have been matched just as well, if not better. Both lived in the same periods between 800 and 600 B.C. a comparatively small difference that the magi in their mysterious ways might have made even less and by the same magi with their wondrous arts these two could have been made almost neighbors, he as a Persian and she as a Chaldean.

I have in mind the case of Semiramis and Zoroaster, both distinctively pagan but each with a likely taste for orthodox religious acceptance had their times been in a different period. Now, we have seen that the pagan doctrines of Semiramis virtually ruled her existence. But soon the currents of Christianity would have reached her watershed and some of that would have been due to the nearby presence of Zoroaster. Since the cult of the Persian prophet was believed at first to be the inspiration of Satan, such a thought would have been outlandish to early Christians. So intense were the feelings aroused by this dark figure from Iran that even Semiramis might have been appalled by his violent gospel. But in the passage of time it became clear he was wrongly judged, and those he described assumed a significance exactly the opposite of his listeners' versions. One scholar, J. Duchesne-Guillemin, in *Everyman's Encyclopedia,* explained the changing roles in this way:

> Although a definite borrowing is still impossible to prove, the resemblances between Zoroastrianism and Judaism are numerous and important.... First, the figure of Satan, originally a servant of God, appointed by Him as His prosecutor, came more and more to resemble Ahriman, the enemy of God. Secondly, the figure of the Messiah, originally a future king of Israel who would save his people from oppression evolved . . . into a universal Saviour very similar to the Iranian Saoshyant. Thirdly, the entities that came to surround Yahweh, such as His Wisdom and His Spirit, are comparable to the archangels escorting Ahura Mazda; the Spirit, in particular is comparable to Spenta Mainyu. The six powers of God are also comparable to the Iranian archangels.

Even at that we have to look elsewhere to find the true saintly measure of this remarkable man, and it seems to be exemplified best in one of the prophet's own quotations unearthed by my friend Kazmer Ujvarosy, a student of Biblical times whose scholarly work, always generously given, has been of great value to me in this and other studies. In *The Secret Books of the Egyptian Gnostics* by Jean Doresse he found this arresting passage on the coming of the Messiah in Zoroaster's words:

Listen, that I may reveal to you the prodigious mystery concerning the great king who must come into the world. At the end of times, at the moment of dissolution which will put an end to them, a child will be conceived and formed with its members in the womb of a virgin, without any man having approached her. . . . He will arise from my family and from my line. I am He and He is in Me. At the manifest commencement of this coming great prodigies will appear in the sky. A star will be seen shining in the midst of the sky: its light will outshine that of the sun. So then, my sons, you are the Seed of Life issuing from the Treasury of the Light and of the Spirit, who have been sown in the soil of fire and of water, you must be on your guard and watch . . . and you will know beforehand of the coming of the great king for whom the captives (of Israel) are waiting to be freed.

Zoroaster's claim that Jesus is from his family and line is significant, Kazmer continued, because it means that both the head of the Babylonian or Persian Magi and Jesus are from this seed of David, and consequently from the 'Seed of the Great Seth.' This suggests that the mysterious Zoroaster is actually Daniel, one of the major prophets. In the *Haggadah* he is described as a scion of the House of David; as a child he was taken to Babylonian captivity and trained with three Hebrew companions who were also of royal seed to serve the king of Babylon; 'in all matters of wisdom and understanding . . .

the king... found him ten times better than all magicians and astrologers that were in all his realm.'

—Daniel 1:20.

Amid the technologies of today we can smile at the Mysteries of Semiramis and the Magic of Zoroaster but can we be certain beyond all doubt that our knowledge in all such matters is greater than theirs? I do not think so, and say that for a reason that revealed itself to me unexpectedly and will, I am certain, stay in my mind for ever.

To retrace events briefly, while publishing *Canadian UFO Report* several years ago I received an article printed in Vol. 3, No. 2, from a young woman in Montana. She was Betty Jones of Kalispell, who became a helpful and diligent correspondent concerning a bell jar shaped object that two picnicking friends of hers had spotted beside a river. They said it was a glowing blue in color and took a picture of the scene which, for some reason when developed, showed no such object. Yet neither the camera nor film was defective.

However, Betty had questioned the couple carefully and had no doubt they were telling the truth.

Thus the bell jar mystery entered my files, with three or four similar stories to follow, one of which involved an air-borne object of the same shape. A high point in the series was reached when Irene Granchi, then our correspondent in Rio de Janeiro, sent us a story on "Antonio's Ordeal." In that article Antonio, whom Irene interviewed in weeks of research, described how he was captured by seemingly mechanical creatures and taken into a huge transparent enclosure large enough to span a playing field that she described as somewhat like a bell jar in shape. Later he was found in a semiconscious state, feverish, incoherent and severely dehydrated, but with no injuries of a kind now associated with mugging. He said his captors showed him strange motion pictures of himself as if he had moved in another dimension.

Later Antonio's condition was considered serious enough to require hospital attention, and more than a month passed before he felt normal again, yet the cause of his illness remained unknown.

A final bell jar report, though brief, seemed to be the most significant of all as it described what may have been this mystifying device in its formative stage. Carried in the Belgian publication *Inforespace* in 1972 and noticed by our alert correspondent Graham Conway, the report followed a remarkable sighting by a pedestrian, Leon Herbosch, in a town near Brussels. At first Herbosch had observed a greenish "blob" on an empty street site. The item continued:

> Within a period of about 30 seconds the blob became brighter, seemed to vibrate, then rose and inflated into a goldish bell shape. The luminous bell enlarged and reached about five meters in height. Its luminosity became almost blinding and its color paler, almost a 'neon white.'
>
> The light, although vibrating, retained the sharp bell shape, its borders being a little like a ray of light very sharp but not limited to a precise place. Its interior seemed made up of tiny luminous particles swarming all around. Once at its maximum height, the object lit up the whole landscape as far as trees a hundred meters behind the site. It looked like a torch of extraordinary power. The whole thing took place in complete silence, nor did the witness notice any heat or smell.
>
> Extremely frightened, the witness saw the bell suddenly 'falling down or going flat' to resume its original shape. The blob then moved away, exactly following the contours of the ground.

That incident, one of the last in my bell jar records, occurred in 1870, still a long way from the era of Semiramis and Zoroaster, but then came the memorable moment when it suddenly became apparent that reports of this mystery did indeed have good reason to be persistent. For behind it was a record going deep into history and touching on the most profound areas of religion and vast arenas of the strange and unexplained.

I learned of this when I opened a book newly published by my friend and colleague, John Robert Colombo, and titled *Mysterious Encounters*. Almost instantly my eye was caught by the words "bell jar" in the middle of one chapter, and in a few seconds for the first time ever I was reading an account of the device that had fascinated me for so long. Also, to my delight, I was also learning there was good reason to be so impressed.

As explained in *Mysterious Encounters* the significance of the bell jar's history lies in its tie with the philosophy of Buddhism and in the magical effects apparently resulting from that association. For examples we have Colombo's permission to reprint passages from the book. These appear in a chapter carried as an interview in 1989 with David Ryshpan who had returned recently from Colombo, Sri Lanka, where a new Buddhist shrine was inaugurated.

We pick up the interview as Ryshpan describes how a High Priest presented him with a relic of the Buddha in the form of a tiny bone. Ryshpan was given the bone as a reward for special services performed in helping to build the shrine. The High Priest handed Ryshpan the small sacred piece in a cloth wrapping. The account continued.

He opened up this piece of cloth and showed me this little piece of bone, which looked like a grain of rice, and he said, "Make sure it is kept under the image of the Buddha in a particular way." Then he explained to me that I should find a fairly heavy wooden board and should then find a glass bell—

A glass bell. . . . ?

. . . a bell jar, the kind you sometimes see in a laboratory where it is used to create a vacuum, and place the relic in the bell jar to make sure it is kept secure so that no one can steal it. I asked, "Why would anyone steal it?" He answered, "Even if someone offered a million dollars for such a relic, they could not obtain one."

I was flabbergasted. Taken aback. The fact was: Here I was, being given such an honor and being shown so much respect by the Buddhist clergy, that they would entrust me with a sacred relic of the Buddha! It was priceless. These relics were handed down from father to son through many generations. They were kept in temples and were always the property of the temple until a new temple was built. Normally they were only taken from whatever reserve they were in and given by one High Priest to another High Priest whenever they inaugurated a new temple. Here all they had was a shrine room.

I assured the Chief Priest I would do my utmost to make sure that it would be kept safe. I appointed my chief gardener, who was a Buddhist, to proceed with making this particular wooden coffer with the bell jar in the prescribed manner. Then it would be placed under the Buddha's image, so that when anyone pays homage to the image they were also worshipping his bones and making various other religious commitments. Two or three times a year the relics have to be paraded before the people on religious festivals like the birthday of the Buddha and the day he attained Enlightenment, etc.

I promised to do this and I made sure that my chief gardener was made the keeper of the shrine. We tried several times to screw bell jars onto the wooden board but the bells always broke. It wasn't until three months had passed that we had the right coffer where we could deposit the sacred relic.

All this time where was it held?

For the whole period of three months it was held in the company safe. Only three people had the combination of the safe. I was one of them. My general manager, who was a Hong Kong Chinese and totally irreligious, was the second. And the third was the accountant, who was a Christian by birth and not really interested in anything. Besides the combination there were the keys to the safe. To open it you still had to use the two keys. The general manager held the keys. Any time the accountant wanted to go into the safe, he would ask the general manager, and it was opened only when the general manager or I was present. For the three months nobody went into the safe without a witness. The whole significance of this is, after three months, when we were ready to deposit the relic in the coffer under the image of the Buddha, we took the brass dagoba or chedi out of the safe and, just to make sure the relic inside was safe, in front of my chief gardener, I proceeded to unfold the cloth which held the relic.

Had you, at any time, or had any one of them, ever checked the relic?

No. It was totally my responsibility, and I left instructions that nobody, except myself, could touch the relic. My general manager assured me that nobody except the accountant ever went into the safe during those months.

The general manager was the witness whenever the safe was opened. To my amazement, when I opened the cloth, there wasn't a relic. There were two relics there.

Identical ones?

No, one was like the original, although I feel that it had changed. It wasn't the same shape. It wasn't any bigger or smaller. It just didn't look the same. And next to it was another relic, not of the same size, but smaller.

What was your reaction?

My reaction was one of shock and amazement. The first thing I did was to run to the telephone and call the Chief Priest and say, "Hey, remember the relic you gave me? Are you sure you didn't give me two because now I have got two."

And his reaction was, "Oh, don't worry. This happens all the time."

"What do you mean this happens all the time?"

He said, "The relics have been known to materialize or dematerialize or travel to other temples or other places of safety."

He said these things are not uncommon? I don't imagine that they happen that commonly. Do they happen that commonly?

He mentioned to me one instance in which there were two temples in adjoining parishes, and they were within about two miles of each other, and one temple had two relics and the other temple had only one relic. There was some animosity that developed between the temples, the reason being that the parishioners felt that at the one with the two relics the priests were not being faithful in their work. They were running little businesses on the side and were being a little bit hypocritical. A year went by. When the elders of the temple inspected the relics, which they did on a yearly basis, the one that had one relic was found to have three relics and the one that had the two relics was found to have nothing. No one would explain it. No one would say why. The feeling was the relics had dematerialized in one temple and then rematerialized in the other temple which was more faithful to their beliefs.

What did he explain had happened in your case?

We were surrounded by three or four temples, and one of their relics could have been transported to our temple. His feeling was that it was very possible that for all the work I had done in building this particular shrine room and in taking an interest in the worker's welfare in a spiritual way, as well as in their physical and mental welfare, had earned me merit. Consequently, we felt that those relics, both of them, really belonged to me, and not to the temple or to the company I worked for. He said if ever I should leave the company or leave the country, for whatever reason, I must take those relics with me and carry them with me at all times.

Well, it so happened that three years later I did leave the company and the country. But before going I went to see him and I said, "I would like to return the relics to you. What do you want to do?"

He said, "No, you are to take the relics. Have a small gold case made for them, and carry them around your neck for the rest of your life. You can leave them in your will to your relatives, or you can will them back to the temple."

I promised to do that. And still, to this day, I carry the relics.

Before they came into your personal possession, where were they kept?

They were kept under the Buddha image for three years.

Was there any reaction from the authorities when you removed them?

No. They felt that the power of the relics had transferred partly into the Buddha image itself. To them the Buddha image is not only sacred it was consecrated. It still retains a certain power from the relics that were kept under it. And they also felt that the relics really belonged to me. The workers themselves never knew about the so-called miracle, as we never told them in order to avoid any hysteria.

The keeper of the temple knew about it and some rather prominent people in government. And every time I went to the Colombo temple, the Chief Priest, now called Loku Hamdru, who had become Chief Priest after the old High Priest had died, used to recount the story of the relics to any visitors who were there at the time, including visiting High Priests of other temples in other countries.

What were their reactions?

Their reaction was one of acceptance that here was a miracle that had happened.

Didn't it bother them that the relic would go to someone who was a non-Buddhist and a foreigner in the country?

It didn't matter. The Chief Priest would tell them about all the work I used to do to the temple, the various donations I made.

A very large rubber plantation of eight hundred acres was willed to the temple. Profits derived from it would be used for Buddhist meditation and spiritual purposes and also to buy books and scriptures. I thought what they were doing was a terrific thing. I went and bought ten thousand clove trees, seedlings, for them. It would take seven or eight years before they would earn any money.

Then they would earn about five million dollars a year from them.

So that was a substantial gift.

It was a gift on my part, but I saw it as a continuing type of thing. Every year there would be substantial revenue to continue their work.

Are they deriving that revenue now?

Yes. It is more than seven years since I donated the trees.

The amazing thing now is to see the Buddhist monks working in a rubber plantation and also setting up daycare centers for the workers' children, food kitchens, growing their own fruits and vegetables. I thought, over all, it was a tremendous thing.

This Chief Priest did tremendous work. I could see that he didn't ask for anything. He used to work sometimes twenty-four hours a day, so I supported him any way I could. His appreciation was shown in giving me these particular relics at the time I built the shrine.

How did you feel when he gave you these things, not being raised in that part of the world, so far removed from that form of culture?

Well, initially, when you think of Buddhism, you think of it as a religion, and usually with religions there are all forms of mysticism and so on. But after talking with various Buddhist monks and with the father of my girlfriend, who had become a monk and who was a very learned person, I learned that it was not a religion but a philosophy. I am very much interested in Buddhism as a philosophy and not as a religion because I am not a religious person. And things like miracles and sacred relics did not register on me.

They were far removed from your own sensibilities?

I have been to various shrines, like the shrine to

Brother Andre in Montreal, with their miracles and holy oil and holy water, etc. I am not saying I am an atheist. I believe in some form of divine being. I can't put my finger on it. But being exposed to the mysticism of the East, initially in Hong Kong, Thailand, Japan, Korea, and India before going to Sri Lanka, and being immersed in the Buddhist culture, I just found it interesting. So when this so-called miracle happened to me, my attitude was, "Okay, so you either believe or you don't."

It happened to me. I have vouched for the fact that it happened. I have witnesses who were there when it happened. And, okay, I laughed it off. It was a miracle. It happened. I can't explain it.

What happens when you tell this story to other Buddhists?

When I tell it to other Buddhists, they say, yes, we can understand, we have heard of similar stories. Whether I tell it Thailand or Sri Lanka, it doesn't matter, they all believe me. It doesn't bother them.

Are they interested in seeing the relic or touching it?

Several people, who claim to have various powers of ESP or clairvoyance, etc., put their hands on it and say, "Yes, tremendous power comes out of it." Well, I take that with a grain of salt.

I can't say it has hurt me. I can't say it has given any benefit to me. All I know is something happened, and I was witness to it. The relics are very precious to me.

When you tell this story to Westerners or to non-Buddhist people, how do they react?

I find that most don't believe in it. The only ones I found who do believe in it happen to be Roman Catholics. They are also supposedly witnesses to various miracles, not personally, but through their indoctrination or the fact that most things could come about because of a miracle. The idea of a miracle is not that foreign to them.

Describe the size of the gold coffer.

It's about three-quarters of an inch long and about a quarter of an inch in diameter.

How large are the relics?

The relics are . . . one would be about, I'd say, a quarter of an inch long and maybe a millimeter wide . . . the other one is about half the size in length but the same width.

You still carry the relics with you all the time around your neck?

Yes.

To gain a fuller understanding of the Buddhist faith, which was now intriguing me more and more I (Magor) bought a copy of *The Stories of the Buddha,* translated and edited by Caroline A. F. Rhys David. Maybe the gods were at hand when I did so because there several more startling lines awaited me which seemed to be an extension of something I had written earlier in discussing the dangers at Chernobyl. At that time I said:

"A spate of UFO incidents in Poland, starting in 1978 and continuing at least into 1983, suggested an eerie possibility that space aliens anticipated disaster in the neighboring Soviet Ukraine, where atomic projects were underway. They may have gathered over Poland to watch results."

Among my informants was the UFO Klub of Wroclaw. It submitted an illustrated clipping giving an artist's version, based on press reports, of elderly farmer Jan Wolski being captured near Lublin by aliens. They were small thin creatures of greenish complexion and a distinct Oriental cast to their features.

Although it seemed like a minor detail at the time, I remembered Wolski heard them saying something to each other that sounded like "TA-TA-TA-TA."

So it surprised me on reading *The Stories of the Buddha* to come across that odd sequence of monosyllables again and to learn from the index that TATA is a Buddhist "masculine appellative of amity." Were Wolski's aliens doing their best to make friends with the old man, identifiable as a paternal figure for whom they might wish to show respect? Also their strange complexions raise thoughts of their possible antiquity. Were they of a bloodline so ancient it far outdated Buddhism itself yet did not discourage them from addressing the venerable stranger in the Buddhist tongue?

Later, showing what may have been their special regard for him, they made gestures encouraging him to undress, which he did, feeling it was wise to obey. They then carefully and gently examined his spare 71-year-old frame.

It was unmistakably a tableau of one society's welcoming curiosity for another from out of their world.

6

One "Enchanted?" Evening

Under some compulsion, which was strange in itself, I started to read articles on the bell jar in its various forms which I had written long before. Suddenly I was looking at one I barely remembered at first. Yet now it struck me as concerning the strangest incident of the lot, and I wondered if for some hidden reason it was now being produced to benefit the total mystery at hand. This was the unchanged story:[1]

> The strange being in the transparent enclosure had his hands up as if to show that he meant no harm. His eyes showed very intense fright.

In those few words William Bosak, seventy, a dairy farmer of Frederic, Wisconsin, expressed an opinion about an "occupant" that may be something entirely different in UFO dialogue. Our visitors often seem shy or

---

[1] "One Enchanted Evening," *Canadian UFO Report*. Vol. 3, No. 5.

elusive, besides appearing inquisitive or even bold at other times, but when Bosak gave me this description of one who was apparently frightened in their unexpected face-to-face encounter, it struck me immediately as unusual.

Are creatures being placed in our midst who are afraid of their strange environment?

Expecting to be laughed at, Bosak kept his experience a secret for several weeks before finding the nerve to tell a local reporter about it. Curious about certain aspects of the story carried in the *St. Paul (Minneapolis) Pioneer Press* that agreed with details of other cases reported in CUFOR, as will be seen, I wrote Bosak for his personal version of what happened. This was his reply:

> I was coming home from a co-op meeting at about 10:30 P.M. It was the early part of December and very mild for that time of year. There were patches of fog on the road so I was driving slowly with my car headlights on low beam.
>
> When I was about one half-mile from home I noticed something in the left-hand lane of the road, so I slowed down nearly to a stop. When I was only a few feet from this object I could see it very plainly. I could only see the top part of the enclosure and the occupant from the waist up.
>
> The strange being in the enclosure had his hands up as if to show he was surrendering or to show that he meant no harm. His eyes showed intense fright.
>
> He was a very strange looking man. The hair on the sides of his head stuck straight out about an inch, but he did not have hair on his face like a beard or whiskers. I tried to see what clothes he was wearing but I believe the fur that covered his body was really his skin and not a fur

suit because I could not see a seam down the front or a collar around his neck like there is on a shirt or jacket. The fur was a sort of reddish brown or a bit darker than what we know as Guernsey color. His ears were more like a calf's ears than like human ears and they stuck straight out at least three inches.

This person was slender and I think that he was about six feet tall but he seemed much taller because I suppose the floor of the enclosure was about two or more feet up off the road. I could not see the lower part because it was in the fog.

He seemed to have quite a flat look to his face, so with the strange ears and hair sticking straight out it made it pretty frightening. Besides he seemed so tall. I also noticed his arms as he stood there with his hands up, and there was hair sticking out on his arms. I could not see any end of a sleeve so I do believe the fur was really his own skin. I did not see his hands very well but he held them right together as he held them up.

When I got right alongside of the vehicle which was about six or eight feet away, he was watching me. As I passed, it seemed the object came right toward my car, and it became very dark in the car. I was looking back as I passed him and I do not know if the car lights dimmed or what caused it to become so dark in the car.

When this object did take off as I passed, there was a sort of swish and it seemed as though something brushed against the car. It did definitely seem as though it came right at me and there seemed to be a tremendous surge of power.

I did not tell you at the start that the first thing I thought of when I saw the object was, now what in the hell is that? I also knew that if I ever told of this, or even dared to tell, I'd have to describe it very thoroughly.

At the end of his account Bosak answered certain specific questions I had asked him in my letter. One of these was to determine what the occupant mostly resembled, man or animal? His answer suggested man.

I could not say what color his eyes were but they were eyes like that of any man. I could not remember the exact shape of his mouth and nose but his neck was moderate in length and like a man's neck. [This seemed to remove him from the Sasquatch-type of creature which has no discernible neck.] His head was about the size of an average man's head.

Asked for more details about the object enclosing the creature and how he fitted into it, Bosak replied:

The transparent enclosure was about six feet across. No, he did not fill the container. He was about six feet tall and slender. The container was standing upright (or hovering?) in the left lane on a blacktop road. There was no snow on the road. I could only see the object above the fog which was about at the occupant's waistline. He did not move toward me but stood there with his hands up and looking possibly as frightened as I was.

Although the creature itself evidently did not move, it appears from Bosak's earlier remark that, as he passed, the enclosure rose with its occupant and almost brushed against the car.

The witness explained that despite a low-lying fog he could see what was happening because the lights of his car illuminated the object. "It did not have any lights of its own that I could see," he added.

Bosak's earlier hesitation about mentioning his experience, and his wish to be thorough when he did so, are understandable. Operator of a 450-acre farm for forty years, he was staking against his story his reputation as a responsible, public-minded member of the Frederic community. The risk was considerable but, as in many other cases, the impact made by his strange sighting was difficult to ignore. Perhaps the real question facing a witness in that predicament is not so much whether to endanger his good name, however distasteful that may be, but whether to be honest with himself by saying something incredible he knows to be true.

Bosak's problem would not necessarily have been easier had he known about other circumstances, but he was not alone in witnessing an incident of that sort. By remarkable chance, just a year earlier Mrs. Reafa Heitfield of Cincinnati, Ohio also had seen a strange creature standing inside a curved glass-like enclosure and she, too, was badly frightened by the unearthly quality of her experience.

Under the heading "Bell jars and Gargoyles" in *Canadian UFO Report* I discussed that and seemingly related cases. This included further study of a photo taken in Idaho by Mr. and Mrs. Orval Wyman of Columbia Falls, Montana, which seemed to show the same kind of transparent enclosure seen by Mrs. Heitfield (her sighting was in October, 1973, and the Wyman photo which was taken in September of the same year) and a sighting by a couple in England five years earlier.

The article concluded:[2]

'... and here attention centers directly on those 'bell jars' which provoke endless questions. Those who believe our visitors are from another dimension may ask if the device provides an entrance between theirs and ours. The extraterrestrialists may ask if, through some magic, it

---

2   Magor, John. "Bell Jars and Gargoyles," *Canadian UFO Report.* Vol. 3, No. 4.

provides immediate transportation to and from other worlds.

'But going far back into time, we come to the most important question of all: Is there a connection between the bell jar device, which apparently is an instant incubator, and the origin of man?'

When I wrote the article, I was not yet aware of the Bosak case in Wisconsin. When that case did come to my attention, it was strangely as if my question was being answered in the affirmative. Naturally that was not so. Nothing in ufology is ever that tidy, and an observer is only fooling himself if he thinks he has stumbled across a neat satisfying answer to anything. For as surely as one day follows another something will happen that refuses to suit his reasoning no matter how he twists it.

But oddly the Bosak incident involved the same type of transparent container seen in the other cases, yet this time the creature concerned was obviously more man-like. Not only did it resemble a man in appearance but its frightened expression suggested an intelligence quite sensitive to the perils of its alien surroundings. We can surmise it was a servant of some sort, probably much inferior to those controlling it, yet possessing a lively awareness of its own. In Mrs. Heitfield's sighting there was a creature that looked and gestured like a ponderous ape. The animal forms in the Wyman photo were not seen in action at all, nor were they even visible when the picture was taken. But on film they have a monkey look much more suggestive of an animal than of man.

So in summary of these three cases, it is conceivable through some long shot guessing that with their bell jar device our visitors can incubate servants or experimental beings of human-like intelligence as well as others of a lower form.

Did they do the same with us in a program for life on earth?

Speculation along these lines inevitably leads to a subject related to the UFO question, that of Sasquatch or the Bigfoot. Observers of the Sasquatch are usually not inclined to see a connection. But to many ufologists, including ourselves, cases in which sightings are adjacent to strange animals forms—and their smells—suggest there is a connection. Some sort of relationship might also explain why ape-like creatures are seen suddenly in places where they have not been seen before, though they may be described in local legend.

Cree Indians have given them names like Weetekow and Saulteaux and the Wendego. But who and what are they?

Obviously, however, it is the Eastern connection established by Colombo's friend David Ryshpan in Chapter 4 to which we must pay first attention. If there is doubt about that, consider the key photo taken years before beside a river in Montana and centered on a glowing blue object (also reported in chapter 4). No one then had the least inkling of an Eastern connection, yet when photographer and artist Brian James developed the picture taken by two sightseers he immediately noticed what he called a "bunch of monkeys" in the background. He numbered them 1, 2 and 3.

It is known that in some regions of the East monkeys are now venerated to a point where there is, for instance, a fable from the city of Benares about monkeys who were put in charge of the King's garden.[3]

---

[3] Rhys David, Caroline A. F., trans. *Stories of the Buddha.*

—By photographer and artist Brian James.

I am sure Brian James in distant Winnipeg was not aware of that when he made his comment. Yet there he was, a professional photographer fully conversant with strange lighting effects, numbering images of monkeys he saw in the photo. Truly we live on the edge of some impossible magic!

Years ago on a hand-written slip of paper I remember coming across what some had obviously intended as a family tree. While this book was still far in the future, all the names were familiar to me except for that of someone called Poland. When I asked June, the family historian who he was, she explained the Polands were close friends of our maternal grandparents, the Fergusons, whose name I carry.

So far so good, though it did seem strange that name should appear on a list with others who were all family members. The mystery deepened when Felicia and I went on what was supposed to be a visit to Hungary but spent most of our time in Poland where we had no known ancestors or relatives, and the mystery burst into full strangeness when we uncovered the remarkable story of Wolski, which follows.

Perhaps it was the quest of that story which drew us in the first place, but considering Hungary was the homeland we so eagerly wanted to see, and did see, that answer is not good enough. My feeling is that the extraordinary attraction of Poland was again the work of family. Spirits. Something happened in Poland, perhaps centuries ago, to which we had to expose ourselves beyond all conscious thought, and we obeyed because other ancestors farther back had spoken.

It was well we did, as it was in Poland we saw the performance of aliens in virtual living clarity.

7

Polish Mysteries

For some reason that beat Ta-Ta-Ta-Ta kept repeating itself in my mind long after I had read of the experience of the farmer Jan Wolski in Poland (chapter 5). That and what I learned then of the rites of Buddhism convinced me that those details had a significance close to the very heart of what Felicia and I were tracking in our eastward odyssey to the mysteries personalized in figures like Semiramis and Zoroaster. Because Wolski, though just a simple farmer, had accidentally made physical contact with aliens, his case quickened our curiosity, despite being far removed in time from the much earlier and far more imposing mystic personalities of history.

But he had something in common with the others, possibly to an even greater degree than they in that his contemporaries believed in the strangeness of his life. Had he not died from old age and perhaps a peasant's poor nourishment as well soon after his remarkable experience, he might have graduated in spite of himself to a more notable place in history.

To learn more about Wolski, Felicia and I paid a visit to an area of Warsaw where some of his friends still lived. Our idea was to meet a few who had known him personally or indirectly. We were happy to find there were several of them, in fact enough to arrange a meeting at the local Hotel Orbis Forum where the staff was quickly able to serve us coffee and I to erect a portable amplifying system my son-in-law had given me as a going away present. From that point, with Felicia serving as "translator" though she speaks no more Polish than I, but accomplishes miracles of meaning with hands and eyes, the meeting succeeded beyond expectation. It made me wonder why no outsider, as far as I knew, had previously gone after the story of Wolski, or in fact the whole story of Poland's exceptional interest in UFO activity. Touching on many aspects of that activity it is nothing short of a capsule classic of a mystery observed in what to us is a foreign country.

In Poland itself, a writer who has pursued the subject diligently is Bronislaw Rzepecki of Cracow and I am grateful to him for use of a book in English he is preparing. It is titled simply, *UFOs in Poland,* and to convey the untouched flavor of his earnest prose, I have not tampered with translation of parts in which he wrote of strange events in Poland at a time when Wolski still lived. At least three of the local reports were picked up by outside agencies. In writing one of these which became a special case, Rzepecki said:

> In 1959, according to witnesses, a huge wheel-shaped object with a fiery smudge and pink edges changing into red flew from the northwest and fell into what is called Polish Basin. It made a rattling sound resembling metal on metal. The water waved rapidly and sprouted to a high level. After several days, harbour guards found on the

beach a being dressed in strange uniform. It spoke an unknown language, part of its face was seriously burned and the figure looked ill and exhausted. Its suit was made of some metal which the hospital service couldn't take off so metal shears had to be used and a bracelet around the waist was taken off.

The strange visitor died soon afterward, and dissection of its body showed substantial difference from humans in location of its organs. The circulation system ran spirally around the whole body.

A second widely reported case in Poland concerned a girl identified only as Emilia W. who went into the woods one morning to pick berries with three other children. There she saw an unusual object with an entrance of four steps which seemed to hypnotize her. Under its influence she mounted the steps and mechanically followed orders from an unseen source. When she was ordered to sit down she obeyed until at some point she became unconscious. In that state she apparently became aware of five small beings in uniform, one of whom she mechanically addressed as "commander." This figure differed from the others in having a hump on its back. None of Emilia's companions were aware of any of this. Afterward the girl remembered she "flew" somewhere but could not elaborate.

A hump-back condition was evident in a third Polish case that attracted special attention. This time, in a sighting that evidently involved an airborne boat, there were three such figures. As Rzepecki wrote:

> Their shoulders seemed to be fitted to their trunks. . . . The second strange thing which surprised the witness—they had very wide hipbones. They looked as if they had no necks and they had humps on their backs where people have their shoulder blades.

They disappeared in a forest. The witness looked around hoping they would appear again. Suddenly he noticed an object over the forest—'one long wall of it.' At that time there were two other witnesses. Both of them noticed the object in the air. It was rectangular with three orange rings on it. It flew up from the lake. After this the dogs who were also there sickened and so were put to sleep.

While some aspects of other sightings would be of interest to ufologists generally, Rzepecki does not tarry long with the commonplace. Instead he moves quickly into a section of reports by Polish pilots which, as far as I can see, is composed of UFO incidents not previously published. One that caught my eye was a report by a fighter pilot at Minsk who recalled an incident at the moment before take-off. "In that instant," he said, "we noticed an object flying over us at about 600 meters. Two streams of flame from jet nozzles were at the back and a bright luminous sheath in front.

It was already dark and there may have been another aircraft there. But we were all shocked because it was extremely close whatever it was, and there was no sound of an engine, although we had heard other sounds from the airfield.

Had the squadron suddenly received a visit from an aircraft with the still unknown features of a Stealth bomber or was it indeed a UFO? A later incident seemed to support the UFO theory as this time, with better visibility, residents of two Polish villages saw a flying object of strange design and behavior. Cigar-shaped with a tapered look at its rear end, it had large windows on its nose and shone with an orange light.

Twice it stopped in midair, an observer reported, then in the next twenty seconds it jumped forward and disappeared over the horizon in a burst of sparks. "All this happened in absolute silence," he remarked, which made it a companion piece of the earlier sighting.

8

Wolski Meets Aliens

This and the following chapter is reprinted entirely from a booklet by Wroclaw's Club of popularization and exploration (Klub Mozaika, ul. Trazemska 2, 53–679, Wroclaw, Poland.) To ensure the author's intent is maintained and to provide insight into Jan Wolski's personality they are carried as originally translated from Hungarian and are presented as published. The only major change is use of italics in the paragraph referring to discovery of a stone which was "an object of worship in ancient times." Having been found in the neighborhood of Jan Wolski's encounter with aliens, this stone may prove of great significance and we suggest a strong effort should be made to give it professional examination.
—John Magor, a Canadian observer.

An incident took place in the village about 140 kilometers southeast of Warsaw in 1978.

It was a cold morning, when a farmer drove through the forest in a cart pulled by a horse. . . . Suddenly he noticed ahead two men going in the same direction. The cart is catching up and passing them. In a moment they are jumping into the cart. In the meantime the farmer noticed, that his passengers, whom he believed to be hunters up till now, had green color of their skin.

The farmer kept on driving along the edge of the forest, in direction of his own meadow. After a while he perceived a white object rising over the meadow, it looked like "a short bus." When the cart was near the white object, the passengers gave a sign to stop. After that they led him to the craft. They got inside using a little lift. There were two other similar looking creatures. On their order the farmer undressed and was "egzamined," after that they let him go away. Using the "lift" he came back to the ground, sat in the cart and drove quickly home to let his family know.

The inhabitants of the village arrived at the glade after a short time but there was no more white object there nor its passengers. Instead they found shoe traces

Jan Wolski

of untypical shape and trails where two persons had walked close to each other.

At the same time a six-year-old boy living a few hundred meters from the meadow saw "a strange plane" flying low over the barn and then rising high up with a roar.

That event took place on Wednesday, May 10, 1978, about 8:00 A.M. in the village of Emilcin in Lublin province. The witness is seventy-one-year-old farmer, Jan Wolski.

Full Particulars of the Incident

That day Jan Wolski, a few minutes to 5:00 A.M., harnessed a four-year-old mare to the cart and drove to Dabrowa Kusmierska. He was taking the short cut as he hadn't managed to shoe the mare and she couldn't walk over the rough surface. There is a road, running parallel to Wolski's, from which it is separated by a narrow strip of the forest and by the fields under crop.

He turned round the corner from a dirt track going into a meadow. The mare was covering a big part of the way for the first time. Driving along Wolski noticed on the right side two persons going in his direction. They were about forty meters ahead. Wolski at first thought they were hunters. He noticed, that they were turning around time after time, as if they were checking their distance from the cart. They may have heard the cart rolling behind.

Wolski noticed the dark green color of their faces. The movement of those two persons you can compare to the walk of a diver on the seas bottom; it was "fluent jumps."

In a place, where there were clumps of trees, Wolski observed that the persons tried to jump over mud and one of them "had done it not very well," meaning that the person made a slip.

When the cart had passed the mud, it got even with the "hunters," who scattered to the left and right, giving a place in the middle.

They walked a while, close to the cart, then jumped in and sat down on both sides. They did it very lightly.

During the drive, the creatures were turning to each other and saying something, but Wolski didn't understand a word of it.

They were still about thirty to forty meters through the brushwood, before they reached the glade. About seventy meters away, a white object hovered in the air near the forest wall.

They covered about forty meters more across the glade when one of the individuals gave a sign with his hand to stop.

The cart stopped near the craft from which slight buzzing sounds were heard.

The horse didn't want to stop at once, so one of the creatures took the driving reins and helped to stop it.

Coming up to the object, the horse was scared and wanted to pass.

The individuals left the cart and gestured to Wolski to do the same. One of them helped him. At that time Wolski felt that they had a very strong hand grasp. Then all three went to the craft. The creatures walked on both sides of Wolski. When they came up to the object, a little lift lowered down on four ropes. One of the strangers stepped into the middle of the platform and motioned Wolski to do the same.

When they had arrived at object, its platform was already on the grass under the object. Wolski had an impression that it was made of wood.

The lift rose up quickly and stopped against the entrance hole. Wolski didn't feel it starting to rise.

In a room with almost black walls there were two other similar strangers. One of them was eating something that looked like an icicle. The other individuals joined the first. The food broke "like bisquits." One of them, by gestures, asked Wolski if he would eat with them, but he refused. Then the person, who had come with Wolski showed him by signs to undress. The individual who had remained on the ground arrived inside, when Wolski was beginning to undress. He took off his coat, smock, sweater and his shirt. The strangers must have been in a hurry, because one of them helped Wolski to unbutton his shirt.

Then he was told to undress completely. When he was doing so, grimaces appeared on the aliens' faces, something like smiles. One of them who was inside, stood in front of Wolski close, and held little plates in his hands.

It wasn't apparent whether the plates were clasped together, or were separate. Their color was grey, they trembled and silently buzzed. It wasn't a metallic sound.

Wolski was standing motionlessly, while the creature with the little plates walked round him and raised his hands. Wolski felt that the creature's fingers were very cold. He also smelt something as if sulphur was burning.

His coat became so saturated with that keen smell after several minutes, that it could be smelt two days later. Wolski tried to conceal the odor from his wife so that she wouldn't suspect it was the cause of her frequent headaches.

According to Wolski, he had been inside the room no more than twelve minutes. When he was undressed, the individuals looked at his clothes, lying on the floor. Especially they were interested in his trouser belt made of leather, buckling and unbuckling it. While doing so, they were saying something to each other.

When they were studying the belt, grimaces appeared on their faces. They looked into Wolski's mouth. Having finished their studies he was told to dress.

The sound close to buzzing, which Wolski had heard outside the craft, was much slighter inside.

Wolski didn't hear inside any footsteps of the creatures. The temperature inside the craft was a little higher than outside.

Standing barefoot on the floor, Wolski felt its coldness. The person, who wasn't directly engaged in examining Wolski, was walking in the room putting from time to time a little black stick, a bar, into two holes in the wall.

It was a bar, the length of a pencil with a diameter of a finger or a pen and ended with a little black knob.

The bar was pushed with a black knob once into one hole, then into the second.

The individual was turning it like a key in the lock, after that he was pulling it out and pushing it into the second hole repeating the same action. In those two holes, two bars were stuck firmly. He pushed two sticks into the holes and moved them.

There were about ten black birds laying on the floor inside the vehicle. They were alive, as they were moving their heads and eyes.

They seemed to be paralysed, they were ravens or rooks. When Wolski put on his clothes, he was motioned to leave. He took off his cap and bowed saying "goodbye," the creatures bowed too and "smiled."

Using the same lift, Wolski found himself back on the ground. The lift stopped over the ground in such a way, that Wolski had to jump down. Jumping out he didn't feel any movement or even swing of the lift platform.

Wolski went up to the cart and turned around. In the exit-hole he caught sight of two or three figures looking out. Coming up to the horse he noticed, that the animal had eaten a great deal of the grass. He sat down in the cart and started. The horse, being afraid of the craft trotted to the road leading to Wolski's house. After 10-15 minutes Wolski was at home. Only his wife was there when Wolski arrived. His sons arrived home after a while. When father had described his adventure, they ran to bring some neighbors, and then went to the glade.

During a visit in Emilcin in June 1978, a friend heard from Wolski's wife, that after returning home "from those beasts her husband had been lying like dead for a couple of hours."

Wolski's sons saw no object in the glade. Looking around they saw traces showing the recent presence of others in that area.

In three places there were well-marked untypical shaped shoeprints. The imprints had a shape close to rectangular. The sole was depressed making a semicircle. Shoeprints were a little longer than the usual length of a man's shoe. Another kind of trace was black feathers cut perfectly even.

There is an information in Nol-Obserwator, that seventy meters east to the place of the light, on the surface of a couple square meters, there was a number of imprints, shoe traces on the grass.

According to later statements of the searchers, it appeared that at a place under the surface, there was a stone which was an object of worship in ancient times, because of the "markings," that covered its surface. With the "passage of time," the stone sank in muddy ground. At present, this stone is in one of the villages near Warsaw.

9

Men or "Strangebeasts?"

Wolski arrived home about 8:00 A.M. In his opinion the event began at about 7:10–7:20 A.M. when he noticed the creatures crossing the meadow.

At the time, there were two children in the farm back yard; six-year-old Adas and four-year-old Agnieszka. The children's mother was preparing breakfast in the house. Between 8:00 A.M. and 9:00 A.M. she heard an immense thunder "was as if it came from underground," being worried about her children she went outside. Seeing her children playing outside, she went back in. After a while her son came and said, that he had seen a plane similar to a bus, flying slowly over the barn. It had one window and he saw a pilot inside the plane. When the plane had passed the yard it rose vertically up and disappeared. At that time a loud thunder was heard.

Look and Behavior of the Creatures

The creatures had a stature of 1.4–1.5 meters, a slim figure, delicately built. They were dressed in an one-piece elastic, black with a shade of grey, rubber-like overalls. The overalls covered the head, remaining uncovered only face and hands, in the down part it passed from knees into shoes.

Dress tightfitting, only close to the waist it was slightly wider. In hadn't any buttons, neither pockets nor belts. Sleeves were fitted in the wrists.

The creatures' heads were average for their size. The face was dark green, like the color of unripe olive, brown green. (Here Wolski was apparently subject to a list of comments.)

Cheekbones set out, giving an Asian character to the face. Slanting eyes elongated horizontally, dark, without the white. In the place of our nose a slight protuberance and two little vertical slits, straight nose. Narrow, straight cut of mouth. According to one questioner, there was a chink in place of a mouth.

Appearing sometimes on their faces grimaces were probably friendly smiles, although they were unpleasant, they consisted in twisting of mouth to one side. In the place of mouth horizonal slit without lips. Teeth white. Face didn't have any hair, hands green, slim and delicate, five-fingered, between fingers there were membranes, reminding one of a blade of grass, surrounding every finger. According to Wolski, fingers, apart from a thumb and a forefinger, connected to the height of the first joint with a thin elastic green membrane. Wolski didn't notice any nails. Their skin was smooth in touch and rather cold. Their hands might have been always cold, or because it was cold outside. There, where man has shoulder blades, they had "a hump" as if under overalls they had rolled a blanket. Wolski stated, that under overalls it wasn't visible if it was the body or any device. Their feet were long and wider, than man's. When the creatures were

From several drawings of UFO Aliens in my files, I believe the one above may resemble more than any others, the type of creature sighted byWolski. Other witnesses have reported seeing a light shining from his belt.

—Composite by Hal Crawford

sitting in the cart, the foot hanging loosely, was bowed in such a way, that it seemed to be a prehensile member. Wolski can't tell the sex. They all were exactly the same. The creatures, in Wolski's opinion, were kind, didn't scream, or push. They behaved very politely and trustingly. They moved with "fluent jumps," you could notice the lightness of their walk.

Their language was quick and fine, for instance "ta-ta-ta-ta. . . . " Their sounds were made in one set without any clear interruption. Wolski stated, they had been talking so quickly, that it seemed to him, they didn't understand each other.

The language you can compare to the sound "bzoo-bzoo-bzoo. . . . " Their speech was described in following way: "a kind of composite barking, croaks and mewing, sometimes a sort of quick snicker or strange bird's twittering."

The pilot in the flying object window "had a green face and looked like a soldier in a grey and black uniform." Behind the glass a brown face and slanting eyes were visible. "He looked like a monkey."

A Friend's View of Wolski

Jan Wolski was born on May 29th, 1907, by profession—a farmer; education—three standards of elementary school. Citizen of Emilcin, the village, in which there are seventy-four farms, no school, no club, no newsstand and only one shop. Till 1978 Wolski in fact traveled nowhere. Before the Second World War he visited the Ukraine then presently the USSR area. His father had been a coachman, so had he. At home he didn't have any radio or TV set. Talking about himself, Wolski maintains that he isn't afraid of anything, he has "cold blood." No nervous reaction during his adventure can be explained in this way: Wolski didn't recognize the situation as a dangerous one. It can be said about Wolski, that he's a solid person, truthful, highly honest, and trustworthy. The citizens of Emilcin maintain that he has never been seen drunk, he doesn't smoke, and hasn't any bad habits. He was bred as a Catholic, treats religion very seriously. He was pleased to swear by God, that everything he says is the truth. He received a letter from a priest with an oath which Wolski signed without hesitation.

Wolski believes, that the persons whom he met were men, but of other nationality, for instance were they Chinese? The color of their skin could have been like that of putting on masks, while the strange look of fingers could be compared to that of wearing gloves. Talking about the creatures, Wolski used words such as: "persons," "individuals," "they" (most frequently), and his wife's word "strangebeasts."

To Wolski the most distinctive personal oddity of all was the odor they left behind. It was not simply a clammy effect that one reads about in ghost stories. It was a keen unpleasant smell that caused Wolski's wife to have headaches and led to her using the name "strangebeasts." If we could identify that smell, we might be on the track of discovering where these aliens come from. As the odor was "keen" and lingering, it may have been sulphurous, prompting us again to think of life underground. Was this an episode like that in chapter 10 concerning the denizens of hell?

10

Hell is for Real!

It is a sort of twisted commentary on the public tastes of our time that while a few professional writers can sometimes wring a fortune from the public with contrived stories related to the fad of UFOs, here was an old man in Poland who died virtually unknown yet he said enough to reveal he had lived through one of the most fascinating of all mystifying experiences. Of course, he did not have an agent, and it almost certainly would have spoiled things for him had that been the case. But the reason for neglecting him was a subtle one that would have had priority anyway. It had something to do with the fact that, obstinately perhaps, we do not believe in zombies or the living dead or ghosts or simply life after death. As in so many cases of the strange and unknown today, the presence of "experts" is needed before a mystifying incident has much chance of acceptance, and even then the chances in its favor are not great.

As an example, I cite an extraordinary case of an item in a highly respected professional publication that did accept a report concerning the supernatural to which its attention was drawn by one of its readers. Yet it was also motivated to explain that U.S. newspapers had refused to print the story.

That publication was none other than the *Biblical Archaeology Review* which carried the story in its November/December issue, 1990, after it was picked up from a prophecy magazine called *The Midnight Cry,* April 1990. The account said in part:

According to a report in a Finnish newspaper scientists drilled a hole in the crust of the earth nine miles deep and accidentally reached hell. The newspaper account quotes scientists saying that when they took soundings for the hole, they heard human screams. Screams have been heard from the condemned souls from earth's deepest hole. Terrified scientists are afraid they have let loose the evil powers of hell up to the earth's surface.'

The manager of the project, undertaken by the Joint European Science Drilling Project, is a Dr. Azzacov who stated: 'The information we are gathering is so surprising that we are sincerely afraid of what we might find down there.'

According to the newspaper report, 'The geologists were dumbfounded. After they had drilled several kilometers through the earth's crust, the drill bit suddenly began to rotate wildly.'

The scientists were surprised not only that 'the deep center of the earth is hollow,' but that it is so hot. According to Dr. Azzacov:

'The calculations indicate the given temperature was about 1,000 degrees Celsius, or over 2,000 degrees Fahrenheit. This is far more than we expected. It seems almost like an inferno of fire is brutally going on in the center of the earth. The discovery of screaming was most shocking to our ears, so much so that the scientists are afraid to continue the project. What we heard turned those logically thinking scientists into trembling ruins. We could hardly believe our own ears. We heard a human voice, screaming in pain. Even though one voice was discernible, we could hear thousands, perhaps millions, in the background, of suffering souls screaming. After this ghastly discovery, about half of the scientists quit because of fear.'

'Hopefully,' Dr. Azzacov added, 'that which is down there will stay there.'

According to a report in the "World Events and Bible Prophecy" section of the April 1990 issue of *The Midnight Cry,* a prophecy magazine, the story is big news in Finland and Norway, whose scientists participated in the project. But American newspapers have refused to print it.

The Midnight Cry story reports that:

The Soviets have canceled the project, have fired the Finnish and Norwegian scientists and have given them "huge financial bribes" to keep silent about the discovery. The Finns and Norwegians took the bribes, we are told, because 'they feared Soviet authorities would execute them on the spot to silence them if they thought they couldn't be trusted to keep silent about this.'

Nevertheless, when these scientists returned home, they divulged these details. As a result, according to *The Midnight Cry,* 'Hardened atheists in Finland and Norway became Christians after realizing Hell is for real!'

In a later issue, bowing to a flock of letters about the item and protesting "We Were Only Kidding," the *Review* apologized for the fuss it had caused and put the blame on a "man from Norway who simply fabricated the whole story." At the same time, referring to unusual news items, it seemed to ridicule any current study of what is thought to be a relic of Noah's Ark. Such an obvious put-down of David Fasold's fine new book *The Ark of Noah* seems unlikely to gain many supporters as it invites criticism of a splendid work which the toughest critic cannot laugh at. Unless, of course, he can establish the Ark never existed, or at least was far too old to have left any trace.

The Review is embarked on a dangerous course if it lets itself be pushed around by a bunch of skeptical readers. This is a humiliation it always used to withstand very well, with a special thought for its spunky combat with meddlers of the Dead Sea Scrolls.

As I approach the point of choosing words with which to bring my story to an end I reflect again that this was the year in which a tragically mistaken figure with Babylonian dreams of destiny bungled history by attempting to terminate civilization as we know it. His only success was that he brought about a climactic encounter between two great civilizations (coincidentally the world's oldest and youngest) though without producing more than a comparatively brief period of self-image reflection for either.

Far down the scale from this major event is the happening I choose for my real ending. It was no more or no less than the strange encounter of Jan Wolski. There are several features of that rare incident that distinguish it from other reports in which aliens have a part.

First, as far as I know, these creatures have never been seen before in such an active and bold group. They behaved without any apparent shyness, making it fairly easy for witnesses to record their actions and number.

Second, they used "props," like the small bar they pushed into a wall. Then there were the little plates, noisemaking plates they held in their hands, and a strange aircraft was seen with what might have been one of their group acting as pilot. Also they had something like biscuits which they ate.

And the stone discovered in the aliens' path may well be the product of events far stranger than we could ever imagine.

Maybe, in fact, that unearthly chain of events had started well before our misguided Babylonian, Saddam Hussein, had made his first big blunder.

In *Canadian UFO Report,* Spring 1979, which I published at the time, there is this eerie report of a UFO activity twelve years before the Gulf War broke out:

> As recently as the last issue we had editorial occasion to discuss UFO interest in the calamities of Iran, man-made and natural. The sign of their interest, by our interpretation, was a dramatic performance over Tehran in 1976 which seemed clairvoyantly to anticipate the civil strife or the earthquake that followed, or both.
>
> Now our visitors have done it again, this time apparently being curious about the world's oil crisis. Last November seven technicians at an oil pumping station near Kuwait in the Persian Gulf watched in alarm and almost disbelief as a strange aerial craft landed in silence nearby. They described it as cylindrical and larger than a jumbo jet. All communications and pumping operations stopped for reasons unknown while it was there.
>
> We can only guess the visitors wanted their craft to be seen because they landed at a busy time and stayed several minutes before shooting up out of sight.
>
> Less than two months later in the Holy Land there were well observed incidents of a different sort. Early one January morning, over Jerusalem's Mount of Olives, UFOs staged a glittering show.
>
> One policeman who drove up for a better look reported:

I saw a startling thing: three objects with irregular hues of red, blue and purple—like a sparkling diamond in the sun. Their size was that of a large street lamp.

Elsewhere over Israel there were other sightings, including one of a red ball over Haifa that was reported over the armed forces' radio. As in the case near Kuwait, the visitors obviously wished to be seen. But this time they were over an area fired up with religious and political strife of world importance.

Clearly the space beings know where to be and when, and just as clearly they want us to know they know.

So the aliens seem to be very much a part of all this. Now, if only we knew more about that stone discovered by Jan Wolski!

> Oh, Semiramis, is it really you who asks me to write
> Of ancestors I never knew,
> The saintly and the wicked,
> The living and the dead,
> But pagans all?
> As the gods I know would not share
> Their secrets with anyone less
> The angel of my heart, dear Semiramis,
> Must indeed be you,
> And I pray you will
> Hear me now.
>
> —John Magor

11

A Pair of Queens

As I have said, the aliens do indeed seem very much a part of this whole story and scarcely were the words off my typewriter when I had further evidence of it. The nearness of their presence was so tangible my hands still tremble from the thought of it. At the very least it meant adding this chapter to my book.

Both my publisher David Hancock and my office manager Barbara Buske, who read every word of what I wrote, will vouch the manuscript had ended with the typing of chapter 10. It was over and done with and ready for whatever publishing plans Hancock might have in mind.

At that point two critical items arrived by mail within less than a week of each other. One was a mimeographed article from Ireneusz Hurij in Wroclaw, Poland, who was carrying on the work of the Wolski story. The other mailing was a finely done book from the Smithsonian Institution titled *Seeds of Change* that we shall look at in a moment.

As if incredibly Semiramis was alerted to my prayer, the article from Poland told me there were not after all any undeciphered markings on the Wolski stone. Instead this block of rough granite had a history which, if anything, made it even more mysterious. It said the first person to take an interest in it was Wolski's father! This happened when he arrived in the village of Emilcin back in the mid-1800s, and the stone even then was in the wooded glade which more than a century later was to be the scene of Wolski's encounter with aliens. And apparently that meeting by itself was not just a matter of chance. For almost from the start, this rough block of granite was known as a "bad stone" or a "devil's stone" and a legend developed that certain unrecorded practices were carried out near it to "turn misfortunes" away.

Before that, however, Wolski's father took a liking to the stone and brought it from the woods as a decoration piece for his house. But soon he was having bad dreams and told friends the stone seemed to be saying "Take me back." Eventually he did so but not to precisely the same spot as before, and the dreams continued. So other moves followed until the stone seemed satisfied and the dreaming stopped.

By that time, however, outsiders had become curious about the mystery, particularly as it was said footprints in the area made strangely dark impressions and cattle that had strayed close to it were sickened. So again the stone was moved, this time for scientific examination (as we had earlier suggested) at a private estate nearby where unfortunately further reports about it have been vague. A credible one, however, is that the object was found to be radioactive, which might explain the bad dreams, to say nothing of the presence of aliens, who were there for the meeting with Wolski.

As I should have realized at once, Semiramis was not reminding me of her presence by forcing a block of granite into my thoughts, though it did indeed answer my curiosity about that matter. What she did instead—or her spirits did—was to tell me more about herself, a subject of which I never tire. She did so by putting an alter ego in her place, and this came from the book *Seeds of Change* that described a woman who in my view personified a near likeness of Semiramis herself. She must have been equally beautiful but was younger by almost 2,000 years! Even so, she was destined to go through the same kind of trials Semiramis had met to find what she considered her mission in life. Also, even as a young woman, she was contending with physical hazards that would have been too much for a brave man, and all this was happening in a developing world foreign to her, just as the expansion into Asia was foreign to Semiramis.

Intrigued, I looked up source material[4] for this counterpart of Semiramis and found it in *The Discovery and Conquest of New Spain* by Bernal Diaz del Castillo. By an early paragraph that caught my eye I became doubly sure I was on the right track. For there it was again—a fearless woman with a male will for achievement.

Receiving the name Marina at a Christian ceremony after she and others were taken as young captives by the Spaniards, she grew up to be a loyal follower of the Spanish Captain Cortes who soon realized what a prize he had in her fluency with the various dialects of Mexico. Her skill in languages helped to make her the constant companion of Cortes for whom it opened many doors to a point where he was called Cortes Malinche. "The reason he received this name," the author explains, "was that Dona Marina was always with him, especially when he was visited by ambassadors or *Caciques,* and she always spoke to them in the Mexican language. So they gave Cortes the name of Marina's Captain, which was shortened to Malinche."

But of course their common interests went far beyond languages, though it took time.

"She was a truly great princess," Diaz says of Marina, "the daughter of Caciques and the mistress of vassals, as was very evident in her appearance. I do not clearly remember the names of all the other distinguished women, but they were the first women of New Spain to become Christians. Cortes gave one of them to each of his captains, and Dona Marina being good-looking, intelligent, and self-assured went to a very grand gentleman who was cousin of the Count of Medellin. And when her husband returned to Spain, Dona Marina lived with Cortes to whom she bore a son."

Obviously Marina had planned everything very well and there I am reminded how Semiramis too planned everything so well. She had a son just when she needed one, and miraculously he was white. But then, of course, Semiramis had that head start of 2,000 years over Marina in acquiring all her know-how.

Even so, Marina gained much of the ground she had yielded to Semiramis for by extraordinary chance her consort, Cortes, was also a master builder, possibly on the same level that Semiramis herself had been.

For evidence of that *Seeds of Change* carries an illustration of a Christian cathedral in Mexico City that Cortes designed to replace the Great Temple dedicated to Aztec Gods. The Gods of Semiramis would have turned in their graves, though it must be said that her own people would soon turn their thoughts to Christianity.

Index

A
Abraham 11
Adapa 57
Adonai 53
Aethiopian 11
Agdistis 28
Ahriman 64
Ahura Mazda 64
Aircraft
  Lancaster Bomber 39-40, 46
  Lysander 37, 39, 46
  Merlin 46
  Stealth Bomber 90
Akkad 27
Alaric, King 58
Allies 39-40
Alpine heights 26
Anatolia 26-27
Andre, Brother 75
Ankara 27-28
  Ankyra 27
Antonio 66-67
Apennine Peninsula 23-24
Aphrodite 16
Arabian Sea 35, 45
Archaeologists 28
Archangel Michael 54
Ark of Noah, The 107
Arpad 57, 62
Asia 9, 17, 50, 56-57, 59, 100, 112
  Central 59
  Minor 27
  Western 31
Assyria
  Great Queen 31
  King 31
Astarte 31
Atargatis 47
Atlantic Ocean 37
Attes 28
Attys 28
Attila 17, 19, 22, 24-25, 58-60
  Scourge of God 22
Avro Canada 40
Aztec Gods 113
Azzacov, Dr. 105-106

B
Babylon 12, 15, 29, 53, 65, 107
  Goddess 25
  Hussein, Saddam 108
  King 29
  Queen 11, 16
Bacchus 25
Bell jar 63, 66-70, 78, 82-83
  glass 69
Belshazzar 29
Benares, City of 84
Bendeguz 17, 19, 22, 58-60, 63
Bennett, Cliff 44
Biblical Archaeology Review 105-107
Bigfoot 84
Black Phaethon 11
Black Sea 27
Bosak, William 78-79, 81-83
Britain 40, 46, 58, 82
British Columbia 21
Brussels 67
Budapest 50, 60-61
Buddha 68-70, 72-73
  Stories of the 76-77

Bulgaria 24
Burgundy 23
Buske
  Barbara 7, 110
  Wayne 7
C
Caciques 113
Caesar, Julius 29
Canada 34, 37
Canadian
  Army 40
  Forces Base 44
  UFO Report 21, 42-43, 55, 66, 79, 82, 108
Carpathian Mountains 22, 60
Cathedral 54-55, 113
Celts 26-28
  The 7, 26
Chaldea 25, 27, 31-32, 63
  Doctrine 13
Chernobyl 76
Christ 28, 53-55
Christianity 25, 28-29, 57, 64, 70, 106, 112-113
Cimmeria 58
Cincinnati, OH 82
Clairvoyance 75
Colombo
  John Robert 7, 68, 84
  Temple 68, 73
Columbia
  Falls, MT 82
  River 21, 25-26
  University 35
Conway, Graham 7, 67
Coptic Texts on Saint Theodore the General 52
Cortes 112-113
  Malinche 112-113
Cracow 88
Crishna 11
Csodaszarvas 52, 55
Csoma, Sandor Korosi 59

Cybele 24-25, 28-29
D
Dabrowa Kusmierska 94
Dagon, mitre of 29
Daniel 65-66
Danube River 9, 22-26, 50, 55, 60
Darius 31
David 65
  House of 65
Dead Sea Scrolls, The 107
Deluge 56
Diaz del Castillo, Bernal 7, 112-113
Discovery and Conquest of Mexico, The 7, 112
Divine 15
  Being 75
  Language 28
  Son 15
Dodwell, Sophie 34
Don River 59
Doresse, Jean 65
Dovaki 11
Duchesne-Guillemin, Prof. J. 64
Dzungaria 59
E
Earth Chronicles, The 7, 30, 40
Earth Station 1 35
Emilcin 94, 98, 102-103, 111
Emilia W. 89
Enki 30, 32-40, 46-47, 56-57, 62
Eridu 35, 40
ESP 75
Esquimalt, BC 44
Eumenes 27
Euphrates 31
  River 27
Europe 39, 59
Everyman's Encyclopedia 64
F
Fasold, David 107

Finland 106
Fire-god 11
Flying saucer 40-41
Frederic, WI 78, 81
G
Galatia 26-27
Gaul 28
Geza
  King 54-55
  Prince 54
Good Master, The 6
Gordon, Crawford 40
Goths 58
Granchi, Irene 66
Gratian 29
Greece 16
Gunderic, King 58
H
Haggadah 65
Haifa 109
Hamdru, Loku 73
Hamilton, ON 33
Hammond, Joanne 42-44
Hancock, David 110
Heaven 52, 54, 63
Heitfield, Mrs. Reafa 82-83
Hell 103-106
Helm, Gerhard 7, 26-27
Herbosch, Leon 67
Hierapolis 47
Himalayas 59
Hislop, Rev. Alexander 7, 10-12, 15-16, 24, 29
Hispania 58
Holy 54
  Land 108
  Oil & water 75
  Trinity 12
Hong Kong 70, 75
Honorius, Flavius 58
Hungar 57
Hungary 6-7, 9, 51, 54, 57, 59, 62

Dzungaria 59
Hongrie 59
Hungaria 59
Uigurs 59
Ungar 59
Hunor 9-10, 50-52, 55, 58, 60
Huns 22, 50
Hurij, Ireneusz 110
Hussein, Saddam 108
I
Idaho 82
India 75
Inforespace 67
Iowa, Battleship 45
Iran 31, 64, 108
Israel 109
J
James, Brian 84-85
Janus, keys of 29
Japan 75
Jerusalem 108
Jesus 55, 65
Joint European Science Drilling Project 105
Jones, Betty 66
Jupiter 28
K
Kalispell, MT 66
Khazars 24
Korea 75
Kush, African Kingdom of 10, 13
Kuwait 108-109
L
Lac Echo 34, 36-37, 40
Lachine
  Rapids 36
  Rowing Club 37
Laurentian Mountains 34-35, 47
Lengyl, Emil 7, 22, 24, 57, 59
Lhasa 59
Lublin 76, 94

M
Madonna 10, 16
Magi 63
  Babylonian 65
  Persian 65
Magor 7, 9-10, 13, 17, 32, 34, 37, 50-52, 55, 60
  Felicia 6-7, 50, 60, 62, 87-88
  John 76, 92, 109
  June 36, 45
  Lillian 7
  Lincoln 7, 40-42, 44-46
  Marion Ferguson 33, 40
  Robert James 30, 33-41, 45-47
Magur 33-34
Magyar 7, 10, 13, 16, 22-24, 50, 54-55, 59
Malton, ON 38-41
Marina 112-113
Medellin, Count of 113
Medes, Indo-Germanic 31-32
Mediterranean 26, 57
Messiah 64-65
Mexico 7, 112
  City 113
Mid-Danubia 24
Middle East 59
Midnight Cry, The 105-106
Milan 58
Mimik 42
Minneapolis, MN 79
Minsk 90
Montana 21, 66, 82, 84
Montgomery, Field Marshal 40-41
Montreal, PQ 33-34, 36, 40, 75
Moon maidens 9-10
Moravia 24
Moses 25
Mount of Olives 108
Mutual UFO Network Journal 7, 46
Mysteries, The 12, 25-26, 31, 66

Mysterious Encounters 7, 68
Mythology 11, 37, 48
N
National Steel Car Co. Ltd. 33
New York Times 45
Nimrod 11-13, 15-16, 60
Nin 12
Ninus 13
  The Age of 11
Noah's Ark 107
Nol-Obserwator 98
North America 21, 35, 106
North Sea 37
Norway 106
O
1,000 Years of Hungary 7, 22, 24, 57
Oregon 21
P
Pacific Ocean 21
Paganism 28
Pergamon 26-27
Pergamos 29
Persia 63-64
Persian Gulf 108
  War 108
Pessinian Temple 28
Peter of
  Chaldea 25
  Galilee 25
Philetairos 26
Picts 25
Pisces 47
Poland 7, 76, 87-89, 92, 104, 110-111
Pope 25
Q
Quebec 36
R
Radium, BC 42
RCAF 37
Rha River 59
Rhea 12

Rhine River 23
Rhys David, Caroline A. F. 76
Rio de Janeiro 66
Roberts, Andy 7, 46
Rocky Mountain Trench 21-22, 26
Roe, A. V. 40
Roman
  Britain 58
  Catholic 75, 103
  Emperor 58
  Empire 28-29, 58
Rome 16, 23-25, 58
  Bishop of 25
  Pagan 25
Ryshpan, David 68, 84
Rzepecki, Bronislaw 88-90

S

Saboath 53
Samsi-Adad 31
San Francisco, CA 51
Saoshyant 64
Sasquatch 81, 84
Satan 64
Saxons 58
Sea Duck 34, 46
Secret Books of the Egyptian Gnostics, The 65
Seeds of Change 110, 112
Semiramis 10-13, 15, 17, 22, 24-27, 30-32, 47, 63-64, 66, 68, 87, 109, 111-113
Seredy, Kate 6-8, 50, 55, 57-58, 61-62
Seth 65
Shem 11
Sitchin, Zecharia 7, 30, 32, 39-40, 47, 55, 57
Smithsonian Institute 110
Soviets 106
Spain 113
  New 113
Spaniards 112

Spenta Mainyu 64
Sri Lanka 68, 75
St. Laszlo 54-55
St. Lawrence River 36
St. Paul Pioneer Press 79
St. Peter 25
St. Theodore 52-54
Stories of the Buddha, The 76, 84
Sumer 27

T

12th Planet, The 32
Tanais River 59
TATA 77
Tehran 108
Thailand 75
Three Tribes 27
Tibet 59
Tigris River 27
Time Chart 30
Tonsure 25
  of Peter 25
Toronto, ON 40
Transmigration 13
Transylvania 26
Two Babylons, The 7, 28-29

U

U. S. Navy 45
UFO 42-44, 51-52, 57, 76, 78, 84, 88, 90, 104, 108
  Brigantia 46
  In Poland 88
  Klub of Wroclaw 7, 76, 92
Ujvarosy, Kazmer 7, 51, 55, 65
Ukraine, Soviet 76, 102
Ural Mountains 60

V

Vac, City of 54
Valley of Gods 42
Vandals 58
Venus 16
Victoria, BC 44
  Flying Club 44

Vienna Gap 22-24, 26
Virgin Mary 55
Vlachs 24
Volga River 59
W
Wars of Gods and Men, The 30, 39, 57
Warsaw 88, 92, 98
Wawzonek 100
West Side Story 61
White Stag 9, 50-52, 56, 59, 61
  The 6-8, 50, 58
Winnipeg, MB 85
Winstedt, E. O. 52
Wisconsin 83

Wolski, Jan 76-77, 87-88, 92, 94-103, 107, 109-111
World War I 37
World War II 46, 102
Wroclaw 92, 110
Wyman, Mr. & Mrs. Orval 82-83
Y
Yahweh 64
Z
Zoroaster 47, 64-65, 68, 87
  Magic of 66

About the Author

John Magor graduated in 1937 with an M.S. degree from the Columbia School of Journalism in New York. He has been an editor for Cavalcade News magazine in London, England, free lance reporter in Washington, D.C., British United Press parliamentary correspondent in Ottawa and publisher of the Prince Rupert Daily News and the Cowichan Leader, both in British Columbia.

In recognition of his work as news correspondent and his "enterprise and leadership" in publishing, Magor received an alumnus award in 1963, from the Columbia journalism school on the occasion of its 50th anniversary.

His contributions to *Canadian UFO Report* were featured in a graphic display at the "Man and His World" exhibition in Montreal. In 1979, after ten years of publishing the magazine, Magor sold his interest to another publisher, but the handwriting was on the wall. A year later CUFOR ceased publication and its loyal little group of volunteers was disbanded.

During World War II John Magor served for five years in the RCAF.

He currently lives with his family in Duncan, British Columbia.